STUDENT STUDY GUIDE

James C. Curl
Modesto Junior College

INTRODUCTORY STATISTICS

Sixth Edition

Prem S. Mann
Eastern Connecticut State University

WILEY

JOHN WILEY & SONS, INC.

Cover photos: Paolo Curto/Getty Images

To order books or for customer service, please call 1-800-CALL-WILEY (225-5945).

ISBN-13 978-0-471-75532-6
ISBN-10 0-471-75532-X

Printed in the United States of America.

10 9 8 7 6 5 4 3 2 1

Printed and bound by Bind-Rite Graphics, Inc.

Preface

Nothing is good or bad but by comparison

– Thomas Fuller

The preface would usually tell you what to expect in the study guide. However an entire chapter, Philosophy of The Study Guide, has been written to explain what is in the study guide and how to use it with Prem Mann's <u>Introductory Statistics, 6/e</u>. I encourage you to take time to read through the first chapter of the study guide to learn about the things that you can do to help you do well in statistics.

This study guide reflects the commitment, hard work, and help of several people. I would like to use the preface to express my thanks to three people who helped me translate the ideas I had about the study guide into reality:

- Brad Wiley II for his vision, encouragement, and guidance.
- Joshua Delahunty, illustrator and typographer, for his untiring commitment and extraordinary talent in desktop publishing in helping produce the study guide.
- Heather Curl, for her ongoing critical reviews and copy editing of the many revisions of the manuscript.

They shared in what has been both a challenging and rewarding experience, creating a unique study guide that is easy to read and use. To the extent that you find the study guide helpful, they each deserve much of the credit.

Finally, I would like to thank the students in my statistics classes, both past and present. Their questions, discussions, and involvement in learning statistics provided the inspiration for this study guide.

Contents

chapter zero

PHILOSOPHY OF THE STUDY GUIDE

*A good notation
has a subtlety and suggestiveness
that at times
makes it like a live teacher.*
— Bertrand Russell

PHILOSOPHY BEHIND THE STUDY GUIDE

The philosophy of this study guide is to help students understand and use inferential statistics. At this point in the study guide, it may be ambitious to expect the reader to feel comfortable with the stated philosophy. It may help to first explain the structure of the text as it relates to inferential statistics. The text is subdivided into three main areas: descriptive statistics, probability, and inferential statistics.

Descriptive statistics is presented in the first section. The calculated values from this section are called sample statistics. The arithmetic mean is a familiar example. The sample statistics are used to challenge accepted or hypothesized values of the population from which the sample was obtained. Presenting and evaluating the challenge is called hypothesis testing and is a major part of the third area.

*Descriptive
Statistics*

The second area of the text involves **probability**. A familiar example from probability involves the chance of getting a head from one toss of a coin. The section on probability introduces concepts that are very important for a better understanding of inferential statistics.

In particular, probability is used:
- To explain how the data we collect is distributed.
- To determine if our sample statistic is a rare event.
- To find the size of the error that can result if we reject the accepted value from the population.

The third area presents **inferential statistics**. Hypothesis testing, a central theme in statistics, is what we typically refer to as inferential statistics. The important point to remember is that inferential statistics incorporates everything we study in the course. It is where all the concepts come together. More importantly, the concepts in inferential statistics are essential to using statistics in applications relating to business, science, medicine, industry, education, research, and many other fields.

The problems students have with statistics often occur at what we will call "Hot Spots". These Hot Spots are concepts that may be difficult to understand or topics that simply require more help than is typically provided in the text. Sometimes it is just a matter of needing more practice so that the concept, formula, or notation is understood. The Hot Spots are designed to provide the additional help and are the focus of this study guide.

One final point needs to be made regarding the philosophy of this study guide. Every effort has been made to make the study guide readable. There is a generous use of space to allow the reader to take notes, respond, and to interact with what has been written. Please do highlight it, mark it up, write in it, and get involved.

THIS STUDY GUIDE ASSUMES

- The focus should be on difficult areas called Hot Spots.
- It is important to understand what you are doing with formulas and why you are using them.
- The use of notation will increase your understanding.
- There is a need to discuss the concepts in a study group to increase understanding.
- You are willing to ask for help from the instructor as needed.

THE CONTENTS OF EACH CHAPTER OF THE STUDY GUIDE

- **INTRODUCTION**
- **HOT SPOTS**
- **REVIEW**
- **DISCUSSION QUESTIONS**
- **TEST**
- **TEST QUESTIONS AND ANSWERS**

OVERVIEW OF THE STUDY GUIDE

Each chapter of the study guide begins with a quotation and an **INTRODUCTION**. The introduction gives a concise overview of the concepts and key terms in the chapter. The content of the chapter is related to the broader picture of inferential statistics that encompasses the subject from the first to last chapter of the text. The important terms of the chapter are highlighted. An important part of the philosophy of the study guide is that the user will become very involved. It is hoped that you will take time on your own, and if possible spend time in your study group, discussing the definitions, formulas, and notation that is part of each chapter.

Introduction

✳
Hot Spots

The next section of the study guide is a list of **HOT SPOTS**. The Hot Spots, the major feature of the study guide, focus on the statistically complex areas of each chapter. Each Hot Spot, designated with the symbol ✳, is keyed to a page in the study guide, and the authors last name, Mann, is used to indicate the related sections of <u>Introductory Statistics, 6/e</u>. You are encouraged to find other Hot Spots and use them as a focus for the discussions in your study group.

Review

The **REVIEW** section consists of the Hot Spots. It is the most important part of the study guide. Each Hot Spot discusses a problem area of statistics and includes examples, and when appropriate, sample problems with complete solutions, and sample problem-and-answer selections that provide additional drill for each problem area.

Discussion Questions

A set of ten **DISCUSSION QUESTIONS** follows the review section. The discussion questions focus on the statistical concepts in each chapter. They can be used as a basis for discussion in your study group or for individual reflection. You are encouraged to meet with your instructor if you have trouble answering the discussion questions.

Test

A ten question **TEST** follows the discussion questions in chapters 2 through 13. The test questions provide additional problems that can be used in preparation for exams. These problems are written with minimal notation, stressing verbal statements that encourage the reader to make decisions about which formula and notation to use.

Test Questions And Answers

The **TEST QUESTIONS AND ANSWERS** section is at the end of the chapter. The test questions are repeated for your convenience, and complete solutions are given for each of the test problems.

HOW TO USE THE STUDY GUIDE

Naturally, each student may use the study guide in a different way. Whatever you do, interact with the study guide. Write in it, add your own personal notes, and make it work for you. The following suggestions may help you make better use of the study guide.

Before the lecture, it is suggested that you read the Introduction section and focus on the key terms that are highlighted. The space in the margins can be used to add notes from your reading of the text and related information from the lecture.

The section on the review of Hot Spots should be used when you are doing homework or are involved in a discussion group. The Hot Spots are listed by title and cross referenced to chapter sections in Introductory Statistics, 6/e. You can also reference a Hot Spot by looking for the Hot Spot symbol, ✳, in the margin of the study guide. The Hot Spots are intended to provide additional explanation of difficult areas. The sample problem-and-solution and problem-and-answer selections that are found in the Hot Spot reviews can be used to both reinforce the ideas presented in the Hot Spot and provide more practice for problems on tests.

Use this margin space as provided to work out your answers to problems in the text.

Although the discussion section is useful in group discussion, it is even more important for individual reflection. Taking time to mentally answer each question provides a good opportunity to see how well the material is understood. If at all possible, try to verbalize the answers to the questions to get an even better feeling for the material.

The chapter test and the related answer section can be used for both test preparation and a general review of the material in the chapter. If more than one chapter is included on an exam, the problems from the chapter tests involved can be divided up to create two or three typical exams. These trial exams can be done in a restricted time frame to provide practice under pressure to prepare for the class exam.

STRATEGIES FOR SUCCESS

Students often want to know if there is some way they can guarantee that they will pass statistics. They want to know what course or courses they should first take, how they should study, what it will take to pass the course, etc. These questions will be answered as you read through this section of the study guide.

The Nature of the Problem

Many authors and instructors alike believe that statistics is not a respecter of background. They argue that both beginning lower division students and graduate students are often equally frustrated when taking statistics. To some extent their statements are valid.

Students with minimal backgrounds in quantitative areas will find the formulas, notation and number crunching uncomfortable. The use of the computer to minimize the trauma of number crunching often creates yet another trauma of learning how to use a computer. On the other hand, the student with a strong math or quantitative background is often frustrated with the language of statistics and the somewhat uncomfortable world of discussion and interpretation.

The reason for the problems most students experience is the dual nature of statistics. The course involves two different areas of thinking. The first involves the verbal area and depends heavily on your ability to learn concepts, discuss ideas, and interpret results. The second involves the quantitative area that depends heavily on notation, formulas, and calculations.

Many students find that they have strong abilities in only one of these two areas. It is rare to find students that are equally strong in both the verbal and quantitative areas. Thus, the math and science students are bothered by the excessive use of language to discuss concepts. The non-math and non-science students get bothered with the formulas and number crunching.

There is a lot that can be done to help what may appear to be a hopeless situation. The following suggestions will help.

Study Groups

The first thing to realize that is you are not alone in your feeling of frustration. There are many other students feeling equally unsure of themselves. But this is the first place you can make a difference. You will find that if you form a small study group so that you can talk not only about your frustration but also about the problems you are having in the course, you will be less frustrated. You will find that you will better understand the material.

The key point is that you must participate in the group. You need to verbalize what you are having trouble understanding and try to explain to others what you understand. You will be amazed at the difference the study group makes. You no longer are sitting alone late at night fighting statistics and getting more and more upset. Instead, you have help if you are stumped. You will also find that you can learn in more depth as you help someone else who is having difficulty understanding. Studying alone usually results in you getting angry with the subject, the instructor, the book, and yourself...not in what should be happening: you learning statistics.

Of course, there is more to success than just forming a study group; but it is an important beginning. As a group you may want to invite your instructor in occasionally to help explain an idea that everyone is having problems with. Your instructor may be able to provide you with some additional sources for problems to do or some practice tests that you can use to prepare for exams. You may want to go together and pay for a tutor to work with you as a group for few hours each week. But most important, you just need the mutual support the group provides as you do homework and prepare for tests.

Before we leave the topic of support groups, it is important to add a comment about how valuable a phone call can be when you are stuck on a problem. A short call to someone in the class can often help you see what you need to do. All you need to do is get the phone number of one or two people in your class and make an agreement as to when you can call each other. It may help to do your homework early enough in the evening so that if you have to make a call, you can do so without waking up the other person.

One of the problems with creating either a study group or a phone network is meeting other students who are in the class. One simple way of dealing with the problem is to prepare a 3×5 card that indicates your interest in forming a study group and lists the times and days you would like to meet. The same could be done with a phone network, and interested students could then meet at the end of a class session to compare interests and time. If nothing else, you could ask the instructor to announce that there is an interest in forming a study group, and for those wanting to get information about the study group to meet briefly after the class.

Computer Readiness

Another thing that you can do to help yourself be successful in statistics is learn how to operate a PC (personal computer) so you can use MINITAB and Excel that are referenced at the end of each chapter of <u>Introductory Statistics, 6/e</u>. Most colleges and universities have introductory courses in the use of a computer. You do not need to learn much more than how to use a mouse and some of the language that seems to be such a part of working with computers. The goal is that you become comfortable with a computer and can use the computer keyboard and mouse. You will be amazed how much just a simple background in typing can help ease the frustration of using a computer. The computer can be a real asset when it comes to actually working with numbers. Your instructor may involve you in a computer lab to do some of your work. The time spent in learning how to use a computer will pay off.

If your statistics course does not involve a computer lab, you will find that it is worthwhile to get MINITAB or Excel and install it on your computer. The MINITAB and Excel instructions that are a part of the text are so complete that you should be able to do the computer related problems that are included in the text. MINITAB and Excel are easy to use and make statistical analysis on the computer a pleasurable experience.

Use Of Calculators

Although you may decide not to use a computer, you really need to invest in a calculator to help minimize the time that you spend doing calculations. There are many calculators available in a variety of price ranges. However, calculators that have a statistical mode are by far the most valuable to you for use in this course.

Although the HP21S is no longer available, it was used for the calculations done in the study guide. <u>Introductory Statisics, 6/e</u> supports the TI83/84 and includes easy to read directions for using the calculator at the end of each chapter. The TI83/84 contain programs that allow you to find both individual and cumulative probability values for the binomial distribution, the value of the correlation coefficient, the equation of the regression line and the predicted values for bivariate data analysis. It also allows you to find both reported P values and critical values for hypothesis tests using the z, t, F, and χ^2 probability distributions.

Whichever calculator you choose, you need to spend some time reading the calculator manual and learning how to use the calculator before starting the course, or at least during the first few weeks. You will then find that it will be much easier to work with the calculator during the semester or quarter.

MINITAB is a registered trademark of Minitab, Inc., 3081 Enterprise Drive, State College, PA 16801, Phone: 814-238-3280; fax: 814-238-4383; telex: 881612.

WHAT COURSES DO I NEED BEFORE STATISTICS?

Most statistics texts assume you have a good background in algebra. It is difficult to survive statistics with less than a background in intermediate algebra. Two years of high school algebra would be the equivalent of what most colleges identify as intermediate algebra. It goes without saying that you need to do well in such courses. A grade of B or higher provides a reasonable guarantee of having a strong enough background. A grade of C, particularly if on the low end of the scale, might not be a strong enough background. However age, maturity, commitment, and good study habits are often as important as the prerequisite course and grades. If you feel you are over your head, take time to visit with your instructor during an office hour. Talk with the instructor about your concerns. The time you spend with the instructor usually provides an answer to the question of whether or not you are prepared to take statistics.

WHAT DO I WANT FROM THE COURSE?

Very often students take statistics because it is a required course for their majors. They simply see the statistics course as another obstacle to overcome on the road to a baccalaureate degree. What most students do not realize is how important the statistics course will become to them if they decide to pursue graduate work. And for those who do not go on to graduate school, many are surprised to learn how they end up using statistics in their job. The point is that you have two choices you can make. You can take statistics as what is often called a "cook-book course" or you can take the course with the goal of understanding statistics.

Students taking the course using the cook-book approach often find themselves overwhelmed with the notation and the number of formulas that must be used and, at times, memorized. Such students do not have any way of organizing the information other than a brute force memory related process. If instead, you approach the course from the perspective of understanding inferential statistics, the formulas can be viewed in a much more constructive manner. For example, the formulas used for what is called a test statistic simply become variations of one simple formula. What was disorganized and overwhelming becomes easier to use and remember.

When the course is viewed from the perspective of understanding inferential statistics, the entire course can make sense and each chapter is a logical extension of the earlier material. You as the student can use this extension as a basis for asking how each new concept adds to the overview of what is being done in statistics. This view of *understanding* statistics, rather than using the cook-book approach is very consistent with the earlier suggestion about getting in a study group to help you not only succeed but *understand* the course itself.

HOW MUCH TIME DO I NEED TO SPEND ON STATISTICS?

Clearly the amount of time each student needs to spend on a statistics course is a very individual decision. The demands of the instructor will be a factor as well as the student's background and general abilities in academia. However, there are some general considerations that are worth discussion.

You must spend enough time to do more than just a "once over" on the problems that have been assigned by the instructor. You need to have time to read the material in the text. There then must be time provided to talk with others in a study group, or anyone else that will listen — even if it is just the wall of your study area. You need to spend time organizing the new formulas and verbalizing the new vocabulary in addition to doing the problem sets assigned. Beyond this, there is the need to study for exams. Assuming that most statistics courses meet three hours a week, the old rule of two hours of study for each hour of lecture seems like a good place to start. Additional hours should be added on for test preparation and study group discussions. A time frame of between 6 and 10 hours per week is realistic in most situations. Obviously, some students may get by with less time, and many students will spend more than this amount of time. A good rule of thumb would be to spend approximately 5 hours minimum during the week on homework and reading and an additional 3 hours each weekend in a study group.

A final point relative to the question of how much time to spend on statistics is in regard to the frequency of how often one should study. There is a great difference in studying one hour each day when compared to doing the five hours in one day. For example, a marathon session on a Saturday will not be as productive as working on statistics on a daily basis.

REVIEWING YOUR MATH BACKGROUND

One of the most important things you can do to help yourself is to make sure you review the material from the algebra courses that you took as a prerequisite for the statistics course. Not everything in these courses needs to be reviewed, but there are some key areas that will help. You do not need to tackle all these areas at one sitting. But as you find yourself getting ready to cover the chapter involved it would pay to take time to review the related material first.

The 20 review areas that follow are related to the specific chapter(s) in Introductory Statistics, 6/e (as indicated with **Stat:**). Chapters referenced in algebra and geometry are also listed (as indicated with **Alg:** or **Geo:**, respectively). Following the 20 review areas are 10 topical reviews that start on page 0–20 and cover many important topics from arithmetic, and beginning and intermediate algebra.

1. **Subscripts** are used with variables, as in x_1, x_2, x_3 to represent different variables in data analysis and in formulas for sample statistics and population parameters.
   Alg: Reference the chapter on sequences and series.
   Stat: Chapter 1 on Introduction...Chapter 12 on Analysis of Variance.

2. The symbol Σ (the **Greek letter sigma**) is used to indicate a sum, as in $\sum_{i=1}^{3} x_i = x_1 + x_2 + x_3$.
   The symbol Σ is used in the definition of sample statistics and population parameters such as the mean and variance.
   Alg: Reference the chapter on sequences and series.
   Stat: Chapter 3 on Numerical Descriptive Measures, ...Chapter 13 on Simple Linear Regression...Chapter 11 on Chi-Square Tests...Chapter 12 on Analysis of Variance.

$\sum x$ *versus*

$\sum x^2$ *versus*

$\left(\sum x\right)^2$

3.   The difference between expressions like $\sum x, \sum x^2$, **and** $\left(\sum x\right)^2$ is important in finding values for sample statistics like the mean, variance, and slope and intercept of the regression line.

Alg: Reference the chapter on sequences and series.

Stat: Used in Chapter 1 on Introduction...Chapter 3 on Numerical Descriptive Measures ...Chapter 5 on Discrete Random Variables And Their Probability Distributions ...Chapter 13 on Simple Linear Regression...Chapter 12 on Analysis of Variance.

Square Versus Square Root

4.   The difference between the **square** (x^2) **and square root** $(\sqrt{x})$ of a number is important to understanding the difference between the standard deviation and variance and the correlation coefficient and the coefficient of determination.

Alg: Reference the chapter on radicals.

Stat: Used in Chapter 3 on Numerical Descriptive Measures ...Chapter 5 on Discrete Random Variables And Their Probability Distributions...Chapter 6 on Continuous Random Variables And The Normal Distribution... Chapter 7 on Sampling Distributions...Chapter 13 on Simple Linear Regression.

Factor Tree

5.   The idea of a **factor tree** to write a composite number as a product of primes is used to create tree diagrams in probability.

Alg: Reference the chapter on least common denominators.

Stat: Used in Chapter 4 on Probability.

A Study Guide For Statistics

6. The idea of a **proportion**; as in a fraction or comparative share of the whole; is used in grouped frequency tables, probability, and hypothesis testing. This proportion is more like a ratio and different from the way a proportion is defined in algebra as the equality of two fractions.

    Alg: Reference the chapter on ratio and proportion.

    Stat: Used in Chapter 2 on Organizing Data...Chapter 4 on Probability...Chapter 8 on Estimation Of The Mean And Proportion... Chapter 9 on Hypothesis Tests About The Mean And Proportion...Chapter 11 on Chi-Square Tests.

    Proportion

7. The meaning of a **variable** and the idea of a **domain** for the variable is used to define probability distributions for both discrete and continuous variables.

    Alg: Reference the chapter on variables and functions.

    Stat: Used in Chapter 5 on Discrete Random Variables And Their Probability Distributions...Chapter 6 on Continuous Random Variables And The Normal Distribution.

    Variable Versus Domain

8. The use of $f(x, y)$ notation to represent a **function** of more than one variable (as in $f(x, y) = 5x^2y^3$) in finding probabilities for the binomial distribution.

    Alg: Reference the chapter on functions and functional notation.

    Stat: Used in Chapter 5 on Discrete Random Variables And Their Probability Distributions.

    Function

9. The idea of an **inequality**, like $x \leq 3$ and the related graph on the number line, is used with function notation to indicate a desired probability with both the binomial and normal distributions.

    Alg: Reference the chapter on linear inequalities.

    Inequality

Stat: Used in Chapter 5 on Discrete Random Variables And Their Probability Distributions...Chapter 6 on Continuous Random Variables And The Normal Distribution.

Pascal's Triangle

10. The expansion of $(p + q)^n$ using **Pascal's triangle** is used to find probabilities for the binomial probability distribution.

Alg: Reference the chapter on the binomial expansion.

Stat: Although Pascal's triangle is not used in the text, it can be used in Chapter 5 on Discrete Random Variables And Their Probability Distributions.

Use Of A Table

11. The **use of a table** to indicate values of a function is used to define discrete probability distributions. The tables are also used to help in the computation of statistics like the mean, variance, correlation coefficient, the slope and intercept of the regression line, and the graph of the regression line.

Alg: Reference the chapter on functions and graphing.

Stat: Used in Chapter 3 on Numerical Descriptive Measures...Chapter 5 on Discrete Random Variables And Their Probability Distributions ...Chapter 13 on Simple Linear Regression...Chapter 12 on Analysis Of Variance.

Area Under A Curve

12. The **area under a curve** is used to find probabilities for continuous variables.

Alg: Reference the chapter on formulas for areas.

Stat: Used in Chapter 6 on Continuous Random Variables And The Normal Distribution.

Transposing A Vertical Axis

13. The concept of **transposing a vertical axis** using the substitution $x = x - \mu$ to create a standard normal curve that is symmetric to the vertical axis.

Alg: Reference the chapters on completing the square.

Stat: Used in Chapter 6 on Continuous Random Variables And The Normal Distribution.

14. The graph of the **exponential function** $f(x) = e^{-x^2}$ is used to define the standard normal curve.

Exponential Function

Alg: Reference the chapter on exponential functions.

Stat: Used in Chapter 5 on Discrete Random Variables And Their Probability Distributions...Chapter 6 on Continuous Random Variables And The Normal Distribution.

15. The idea of **symmetry of a curve** like $y = x^2$ is used to simplify using a table .

Symmetry Of A Curve

Alg: Reference the chapter on graphing second degree functions in two variables .

Stat: Used in Chapter 6 on Continuous Random Variables And The Normal Distribution.

16. The graph of the **linear function** $y = mx + b$ using the slope intercept form of graphing is used in bivariate data analysis.

Linear Function

Alg: Reference the chapter on graphing linear equations.

Stat: Used in Chapter 13 on Simple Linear Regression.

17. **Greek letters** are used in statistics to represent special values. The letters α (alpha) and β (beta) are used to represent probabilities in hypothesis testing and the letters μ (mu) and σ (sigma) are used to represent values of the population parameters.

Greek Letters

Geo: Reference the chapter that introduces notation.

Stat: Used in Chapter 3 on Numerical Descriptive Measures ...Chapter 9 on Hypothesis Tests About The Mean And Proportion.

Hypothesis

18. The idea of a **hypothesis** from geometry is used to create a null and alternate hypothesis in hypothesis testing.
    Geo: Reference chapter on creating hypotheses.
    Stat: Used in Chapter 9 on Hypothesis Tests About The Mean And Proportion.

Indirect Proofs

19. The idea of proving something in statistics is very similar to **indirect proofs** in Geometry where the assumed hypothesis leads to a contradiction and is thus rejected in favor of the alternate position. In statistics if the assumed hypothesis leads to a rare event, then the hypothesis is rejected in favor of the alternate.
    Geo: Reference the chapter on the logic of proofs.
    Stat: Used in Chapter 9 on Hypothesis Tests About The Mean And Proportion.

False Hypothesis

20. In Geometry, **if the hypothesis leads to a false result**, the hypothesis is wrong. In statistics, if the hypothesis leads to an unlikely result, the hypothesis is probably wrong.
    Geo: Reference the chapter on hypothesis testing and the logic of proofs.
    Stat: Used in all the chapters on hypothesis testing.

Although the prerequisite for most statistics courses is a minimum of one year of algebra and often stated as the equivalent of an intermediate algebra course, most students find that they have forgotten much of the content of the prerequisite courses. The intent of this section is to present a review of a few topics from both arithmetic and algebra that are essential to your understanding of statistics. It isn't necessary that you recall everything from the arithmetic and algebra courses you have taken. In fact, there are relatively few concepts that you will need to review; but it is important that you understand these few ideas very well. The list of specific review topics includes the following ten topics:

1. Using the Equal Sign Correctly.   **page 0–20**

2. Changing From Percents to Decimals.   **page 0–21**

3. Using the Radical Sign.   **page 0–22**

4. Creating Intervals.   **page 0–23**

5. Graphing the Equation of a Straight Line.   **pages 0–23, 24, 25**

6. Graphing Exponential Curves.   **pages 0–26, 27, 28**

7. Symmetry .   **pages 0–29, 30**

8. Using the Binomial Expansion.   **pages 0–31, 32**

9. Using Factorials.   **pages 0–33, 34**

10. Using Sigma Notation.   **pages 0–35, 36**

One final point needs to be addressed before we review the above topics. It is very important to develop a number sense and be able to do simple computations in your head with out the use of a calculator. The ability to do mental arithmetic is very important as concepts are discussed in a statistics lecture. It is true that both calculators and computers are used to work with large data sets, but it is important to work with small values in your head.

As you go through the following review you may find topics that you need to spend more time on. You might want to get some help from your instructor or a tutor so that you understand the material in the review. You may find that referencing the text you used in the course you took in mathematics before taking statistics will help. At any rate, it is important that you take the time to understand each topic area before you need it in the statistics course you are taking.

Using the Equal Sign Correctly:

There is a common error in using the sign of equality (=) that relates to using a table in statistics. All of us are familiar with the process of using a table to find a square root. For example, if we do not have a calculator handy and we look at a table to get the square root of 10, we find the value given as 3.16 to the nearest hundredth. We then correctly write that $\sqrt{10} = 3.16$ for the approximation.

A similar use of tables in statistics allows us to find the area under a curve to the left of a value we call z. The symmetrical bell shaped curve that we find in statistics looks like the following where the area shaded to the left of 0 represents 50% of the area under the curve. A z-table in statistics can be used to find the area under the bell shaped curve for different values of z. If z=0, the table gives the value of 0.5 meaning that area to the left of the vertical line z=0 is 50%. Students often incorrectly write that z=0=0.5 which is clearly an error since $0 \neq 0.5$ **[ERROR]**.

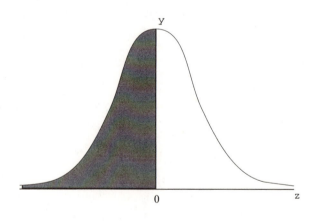

We get around this problem of the incorrect use of the equal sign by using notation that tells us we want the area to the left of 0. We tell the reader that we are interested in the region to the left of 0 by writing $z < 0$. We can then write $A(z < 0)$ to represent the area to the left of 0.

The area to left of the vertical line at 0 is interpreted as being the same as the chance or probability of getting a value of z in the region defined by $z < 0$. Rather than using $A(z < 0)$, we write $P(z < 0)$. The correct way of telling the reader that half the area is to the left of 0 is to write $P(z < 0) = 0.5$

Changing From Percents to Decimals:

Although it may seem trivial to review converting from a percent to a decimal, beginning stat students often make careless mistakes when they make the conversions. The rule for changing from a percent to a decimal is as follows:

To change from a percent to a decimal, move the decimal point two places to the left and drop the %sign.

For example:
95% becomes 0.95 and
1% becomes 0.01 and
2.5% becomes 0.025

If we remember that percent means over one hundred, then an expression like 1% should be interpreted as $\dfrac{1}{100}$ and written as 0.01 when we want the decimal. A common error in statistics is for students to write 1% as 0.10 [**ERROR**]. If we understand that the symbol for percent, %, means over one hundred, we should be able to avoid mistakes like saying 1% is 0.10 [**ERROR**]. This type of error is serious in what we will do in tests of hypotheses and writing confidence intervals in statistics.

Using the Radical Sign:

A minor problem occurs using square roots. Students often only focus on the first word of the phrase "square root" and incorrectly square the value in question. For example, if we want the square root of 9 we should get the answer 3. But if we only hear the word square, we end up with the wrong answer of 81 [**ERROR**].

Fortunately, the radical sign, $\sqrt{}$, is used in statistics to indicate that we are to take the square root and we can match up the symbol on the calculator with the symbol in the expression we have in front of us.

There is still a possible glitch in that both the "square root" and "square" key are on the same key on some calculators. When this happens one must hit a second key to get to the function desired. The point is that care must be taken to make sure that you are getting the square root correctly. This error becomes very serious when we find what is called the standard deviation in descriptive statistics.

Creating Intervals:

Intervals in algebra are often given as an inequality such as $1 < x < 3$ and the graph on the number line is shown as follows where the open dots indicate that x does not include either the 1 or the 3 :

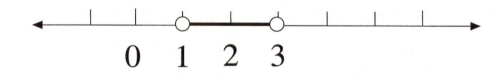

Sometimes we generate such intervals using an expression like 2 ± 1. The $2 + 1 = 3$ gives us the right side of the interval. The $2 - 1 = 1$ gives us the left side of the interval. We write the open interval as (1,3). Note, that the smaller value is written first followed by the larger value with a comma between the two numbers. We call such intervals closed intervals if the end points are included. We will create closed intervals in statistics when we create what are called confidence intervals to estimate the center of a population distribution.

You might recall that you obtained similar expressions when we used the quadratic formula to solve 2nd degree inequalities of the form $ax^2 + bx + c \leq 0$. We first solved for equality using the quadratic formula

$$x = \frac{-b \pm \sqrt{b^2 - 4ac}}{2a}$$

For example, if we had $x^2 - 4x + 2 \leq 0$, we would have a=1, b=-4, and c=2.

$$x = \frac{4 \pm \sqrt{8}}{2} = 2 \pm \sqrt{2}$$

We then get $x = 2 + \sqrt{2} = 3.4$ and $x = 2 - \sqrt{2} = 0.6$ as the end points of our interval. We would finally write the solution to the quadratic inequality as $0.6 \leq x \leq 3.4$ or [0.6,3.4].

Your ability to create an interval will be very important in statistics when you create confidence intervals and when you use the Empirical Rule to predict the amount of data in a particular interval of a bell shaped distribution.

Graphing and Writing the Equation of a Straight Line:

One of the very important skills you learned in a beginning algebra class was to graph points on a rectangular co-ordinate system. Knowing how to plot ordered pairs using the x,y axes provided the background to graph lines and curves. In fact, we defined equations of the form $ax + by = c$ as linear equations because the graph of such equations are lines.

For example the following set of ordered pairs satisfy the equation $2x + 3y = 6$:

$$\begin{array}{c|ccc} x & 0 & -3 & 6 \\ \hline y & 2 & 4 & -2 \end{array}$$

When we graph the set of ordered pairs we get the following graph:

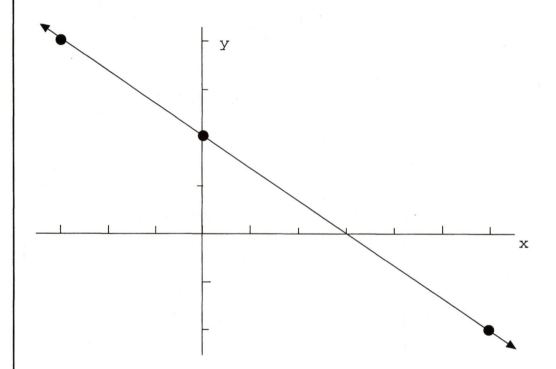

There are two other methods for graphing a line that you should have learned in an algebra class. One is called the two intercept method and the other is called the slope intercept method.

Using the two intercept method we let $x=0$ and solve for y and then let $y=0$ and solve for x.

When $x=0$, we get $y=2$ for the ordered pair (0,2).

When $y=0$, we get $x=3$ for the ordered pair (3,0).

When we plot the two intercepts on the coordinate system we again get the same line that is graphed above.

The slope intercept form requires that we first solve the linear equation for y. The coefficient of x will be our slope while the constant will be our intercept on the vertical axis. In our example of $2x+3y=6$ we solve for y and get the following: $y = \dfrac{-2}{3}x + 2$ where

the slope is $\dfrac{-2}{3}$ and

the y intercept is 2.

We start by locating the intercept (0,2) on the vertical axis. We then go three units to the right and down two units to get to the second point on our graph. We join the y intercept with our second point to get the graph of our line.

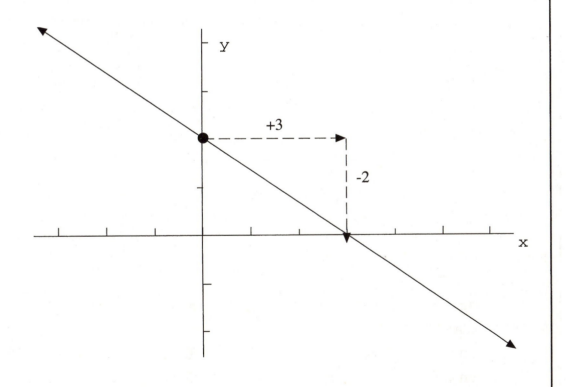

You can read more about the slope intercept form of the equation of a line in Chapter 13, Hot Spots 1 and 2 of this Study Guide.

Graphing Exponential Curves:

The idea of a bell shaped curve is very important in statistics. The bell shaped curve is defined using an exponential function that is usually learned in an intermediate algebra course. The simplest exponential function that is found in algebra is given by the equation $y = 2^x$. Note that this is a different function than $y = x^2$ that generates the curve called a parabola that we see in suspension bridges such as the Golden Gate bridge in San Francisco.

We get the exponential function changing the base and exponent to get $y = 2^x$. When we create the table of ordered pairs for the equation $y = 2^x$, we get the following:

$$\begin{array}{c|ccccccc} x & -3 & -2 & -1 & 0 & 1 & 2 & 3 \\ \hline y & \frac{1}{8} & \frac{1}{4} & \frac{1}{2} & 1 & 2 & 4 & 8 \end{array}$$

When we graph the table of ordered pairs, we get the following curve:

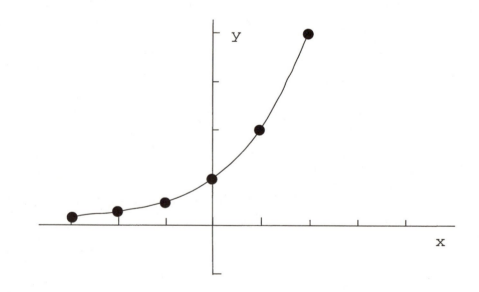

The usual intermediate algebra presentation then involves changing from a base of 2 to a base called e, the base of the natural logarithm, (Note: $e \approx 2.71828^+$). Our new equation is $y = e^x$ and getting a table of ordered pairs for this graph requires that we use a calculator or a table. Regardless of which is used the graph looks very similar to $y = 2^x$.

Unfortunately, many of the intermediate algebra presentations do not go much beyond the graph of $y = e^x$. However there are two more exponential functions that we can graph that are closely related to the bell shaped curve. The first is $y = e^{-x}$ and the second is $y = e^{-x^2}$.

It is easier to work with $y = 2^{-x}$ and $y = 2^{-x^2}$ to see the changes that occur with the new functions. We will start with the graph of the table of ordered pairs for $y = 2^{-x}$;

x	−3	−2	−1	0	1	2	3
y	8	4	2	1	$\frac{1}{2}$	$\frac{1}{4}$	$\frac{1}{8}$

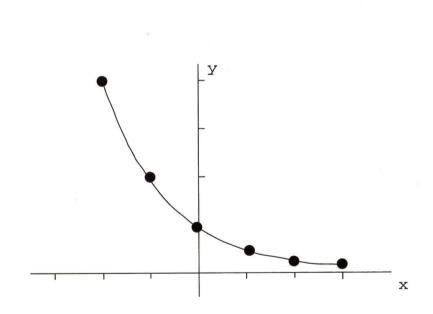

Now we graph the table of ordered pairs for $y = 2^{-x^2}$

$$\begin{array}{c|ccccccc} x & -3 & -2 & -1 & 0 & 1 & 2 & 3 \\ \hline y & \frac{1}{8} & \frac{1}{4} & \frac{1}{2} & 1 & \frac{1}{2} & \frac{1}{4} & \frac{1}{8} \end{array}$$

If we join the points with a smooth curve we get a bell shaped curve like the bell shaped curve we use in statistics. In statistics the bell shaped curve is actually given by the equation $y = e^{-x^2}$.

As a footnote it should be pointed out that we want the area under the bell shaped curve to equal one and have approximately 68% of the area under the curve between the points where the curvature changes (between $x = -1$ and $x = 1$ in the above graph). We accomplish this by dividing the right side of the equation by $\sqrt{2\pi}$ and using an exponent of $\dfrac{-x^2}{2}$; so the equation is written as

$$y = \frac{1}{\sqrt{2\pi}} e^{-\frac{x^2}{2}}.$$

We call the curve given by this exponential function the density function of the normal distribution that we use in statistics.

Symmetry:

The idea of symmetry is very useful when we are sketching the graph of a function. For example the equation of the parabola $y = x^2$ can be graphed by using only positive values of x. Since $f(x) = f(-x)$ we know that the function is symmetric to the vertical axis (y axis). Once we plot the points in the first quadrant where both x and y are positive, we can simply sketch in the left side of the graph knowing that the function and graph are symmetric to the y axis. The graph below shows the right side of the curve for positive values of x.

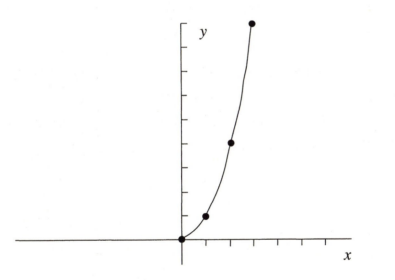

Now, using the idea of symmetry we can graph the left side of the curve. The easiest way of doing this is to draw horizontal line segments going to the left of each point we have graphed so that the y axis bisects the line segments. Then we simply join the end points of the line segments as shown in the diagram on the next page. This has been done with a dashed curve to show how the left end points of the line segments have been joined. When we have symmetry to the vertical axis we can fold the graph on the y axis and the two parts of the graph will be on top of each other.

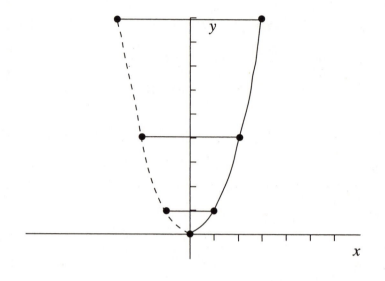

The idea of symmetry is very important in statistics as the bell shaped curve is symmetric to the vertical axis. We use this idea to help us find values on the horizontal axis that give the same area in the tail of curve. In the curve below 5% of the area is to right of $z = 1.645$.

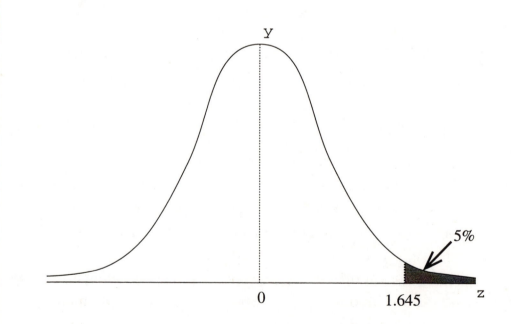

Because the bell shaped curve is symmetric to the y axis, we know that 5% of the area under the curve will be to the left of $z = -1.645$. We use what is called a z table to get one of the two values of z and get the second value of z using the idea of symmetry. The graph below shows the 5% area to the left of $z = -1.645$.

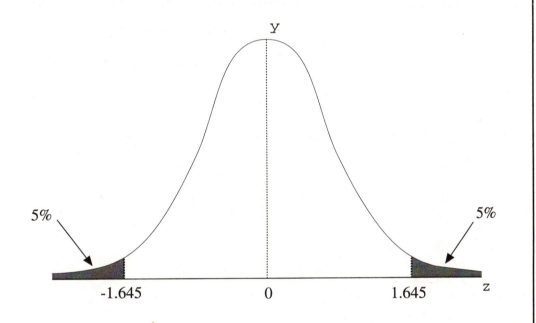

When we look at the shape of distributions in statistics one of the first observations we make is whether or not the distribution is symmetric. We also will use the idea of symmetry to find values from tables such as was illustrated above.

Using the Binomial Expansion:

One of the important skills that comes from beginning algebra is the expansion of a binomial. This is first done with the FOIL method that is used to find the product of binomials. The short cut for squaring a binomial is to square the first term; double the product of the two terms , and add on the square of the last term to get

$$(p+q)^2 = p^2 + 2pq + q^2.$$

The next step in a beginning algebra course is to learn the expansion of the third power of a binomial. Although this can be done by multiplying $p^2 + 2pq + q^2$ by $(p+q)$, the product is usually done by memory and then practiced in problems involving factoring.

$$(p+q)^3 = p^3 + 3p^2q + 3pq^2 + q^3$$

Finding the binomial expansion for exponents of 4 or more is a topic in intermediate algebra using the formula

$$(p+q)^n = \sum \binom{n}{k} p^k q^{n-k}.$$

Another approach to finding the coefficients of the binomial expansion is to use Pascal's Triangle:

```
                1
             1     1
          1     2     1
       1     3     3     1
```

The triangle is formed by using the coefficients of the terms of the binomial expansions. For example the row 1 3 3 1 refers to the coefficients of the expansion of $(p+q)^3$ that is given above. Using Pascal's triangle is reviewed in this Study Guide in Chapter 5 in Hot Spot 5. If you are rusty with binomial expansions you should take some time to review the topic and practice before you do the material on Discrete Random Variables and Their Probability Distributions.

Using Factorials:

The binomial expansion that is learned in both beginning and intermediate algebra is very useful in statistics. We use the binomial expansion to find the chance of certain events occurring. For example, we want to find the chance of getting 3 heads in the toss of 5 fair coins, we can use the binomial expansion $(p+q)^5$ to get the answer. The answer to the question is found in the term that has the product p^3q^2. In order to find the coefficient of this term we can use a formula for the combination

$$\binom{5}{3}$$

This symbol is called a combinatorial and refers to how many ways we can select three things from five elements. The formula for calculating the number of combinations that are possible involves what we call a factorial. The definition of n factorial, written as n!, involves the following product:

$$n(n-1)(n-2)\cdots(3)(2)(1).$$

If we want $\binom{5}{3}$, we find the following quotient: $\dfrac{5!}{3!(5-3)!}$

where $5! = 5 \times 4 \times 3 \times 2 \times 1 = 120$
and $3! = 3 \times 2 \times 1 = 6$
and $2! = 2 \times 1 = 2$.

$$\binom{5}{3} = \frac{5!}{3!\,2!} = \frac{120}{6 \times 2} = 10$$

We now have the coefficient of the product p^3q^2 and our answer to the original question is $10\,p^3q^2$, where p = the probability of getting a head and q = the probability of getting a tail on the coins tossed. Both p and q equal 0.5 for a coin so the answer to the question is $\dfrac{10}{32}$.

A couple of additional points need to be made. Factorials can get large very quickly. For example 20! is a very large number and using a calculator the answer is given as 2.432702×10^{18}. Because of the size of factorials we often use a calculator to find their values, but we can use a shortcut for smaller factorials. For example if we wanted to find $\binom{20}{15}$ we could write the solution as follows:

$$\binom{20}{15} = \frac{20!}{15!\,5!} = \frac{20 \times 19 \times 18 \times 17 \times 16 \times 15!}{15! \times 5 \times 4 \times 3 \times 2 \times 1}$$

We can then cancel the common 15! and cancel the remaining factors in the denominator to get the product

$$\binom{20}{15} = \frac{19 \times 6 \times 17 \times 8}{1} = 15{,}504.$$

Although the answer is a large number, the arithmetic was manageable. The short cut of writing the numerator as a partial product times a factorial is a useful process that makes it possible to work with large combinatorials.

One last point. Some times we need to deal with the special factorial, zero factorial. We define zero factorial to be one and write the following expression:

$$0! = 1$$

You can read more about using factorials to find probabilities like the one in our example in Chapter 5, Hot Spot 4 of this Study Guide.

Using Sigma Notation:

The symbol $\sum$ is called sigma and is used in statistics to define some of the concepts that are used. $\sum$ is the same symbol we used in intermediate algebra to write a sum. For example, if we want to write the sum $1+2+3+4+5$, we can use sigma notation to indicate that we want to add up the positive integers between 1 and 5. We would write

$$\sum_{i=1}^{5} i = 1+2+3+4+5.$$

The expression tells us to sum the letter i starting with 1, increment by 1 each time, and go up to 5. The letter i is used as an index and the $i=1$ to 5 tells us where to start and where to end.

As a second example if we wanted to add up the squares of the first five positive integers we $1^2 + 2^2 + 3^2 + 4^2 + 5^2$ we would write

$$\sum_{i=1}^{5} i^2 = 1^2 + 2^2 + 3^2 + 4^2 + 5^2.$$

The expression tells us to start with 1^2, then add in 2^2, and continue until we reach 5^2. In this example the index, i, is squared each time.

In more general terms the index, i, is used along with a variable, x, to create what we call a subscripted variable, x_i. We use subscripted variables to represent the numbers in a data set. For example, if we have the five test scores 75, 86, 83, 94, 68, we can use the subscripted variable as follows:

$$x_1 = 75$$
$$x_2 = 86$$
$$x_3 = 83$$
$$x_4 = 94$$
$$x_5 = 68$$

If we want to write the sum of the five test scores, we can use sigma notation and write $\sum\limits_{i=1}^{5} x_i$ which tells us that we want the sum $x_1 + x_2 + x_3 + x_4 + x_5$. Again the idea is that the index i starts at 1. We know this because $i=1$ is at the bottom of the sigma symbol $\sum$. We are to go up (increment) one each time we write x_i until we get to the top number 5.

We start with x_1 (read as x sub one) when $i=1$. We then increase the index by one to get $i=2$ and write x_2 (read as x sub two) which we add to x_1 to get $x_1 + x_2$.

Now we increase $i=2$ by one to get $i=3$ and write x_3 (read as x sub three) which we add to what we already have to get $x_1 + x_2 + x_3$. We continue incrementing two more times until $i=5$ and have the sum $x_1 + x_2 + x_3 + x_4 + x_5$.

$$\sum_{i=1}^{5} x_i = x_1 + x_2 + x_3 + x_4 + x_5$$

In recent years it has become more common to write the expression $\sum\limits_{i=1}^{5} x_i$ simply as $\sum x$ to indicate we are to add up all of the values in our data set that are represented by x. The popular wisdom is that students taking statistics will be more intimidated by writing the sigma notation with the index i in the expression.

You can read more about sigma notation in Chapter 3, Hot Spot 3 of this Study Guide. You might also now note the symbol $\sum\circ$ used in the Study Guide to indicate the end of a chapter.

$$\sum\circ$$

chapter one

INTRODUCTION

*Some people hate
the very name of statistics,
but I find them
full of beauty and interest.*

– Sir Francis Galton

INTRODUCTION

We start our study of **statistics** in this chapter with several definitions. The definition of statistics is divided into two areas, descriptive and inferential. We define **descriptive statistics** as the branch that includes collecting, organizing, displaying, and describing data. In contrast, **inferential statistics** is defined as the branch that uses sample results to help make decisions or predictions about populations.

We will be working with **data** that we collect from what is called a **population**. The population consists of all elements whose characteristics are being studied, and the portion of the population we select for study is called a **sample**. The sample is simply a collection of some of the data from the population. The data itself can either be quantitative (can assume any numerical value) or qualitative (can not assume a numerical value.

We will use the sample data to calculate a statistic. A **statistic** is defined as a numerical property of the sample. The test average that we determine in each of our courses at the end of a quarter or semester is a familiar example of a statistic. The sample statistic is used as an estimate of a population **parameter**, a numerical property of the population. The accuracy of our estimate depends on how well the sample represents the population. As such we are concerned that a sample is a **representative sample** that contains the characteristics of the population.

Both the sample statistic and the population parameter become an important focus in inferential statistics. Generally speaking, we use the sample statistic to challenge an accepted value of the population parameter. If the statistic and parameter differ greatly, we argue that the assumed value of the parameter should be rejected. Much of our work in the following chapters is an effort to quantify the phrase "differ greatly". Again, our decision will depend on how representative the sample is of the population .

There are several methods used to obtain samples. The method that is important to later work is called a **random sampling**. If each element of the sample has the same chance of being selected, we call the sample a **simple random sample**.

At times, there are limitations on the resources available for obtaining a simple random sample. We then may have to choose between several other sampling methods that involve a chance process to get the sample, but do not qualify as simple random samples. The stratified sample, cluster sample, and systematic sample are then used. The **stratified sample** involves dividing the population into groups we call strata; we then take a sample from each strata. The **systematic sample** involves taking every nth piece of data from a list of the population; whereas, the **cluster sample** involves taking a random sample from some selected group (cluster) of the population.

There are two common **non random samples**. The first is called a **sample of convenience**; it is a sample that already exists and is available for study. The second is a **judgement sample** created by an expert. In both cases the elements are not chosen by a chance process.

CHAPTER 1 HOT SPOTS

1. **Descriptive Statistics Versus Inferential Statistics.**
   Starts on **page 1-4.** Mann 1.2.1, 1.2.2

2. **A Sample Versus A Population.**
   Starts on **page 1-5.** Mann 1.3

3. **Hot Spot #3 – $\sum x^2$ Versus $\left(\sum x\right)^2$**
   Starts on **page 1-6.** Problems on **pages 1–7, 1–8.** Mann 1.8

If you find other HOT SPOTS, write them down and use them as a focus of your discussions in the study group. Or you can use the HOT SPOT as a the topic for a help session with your professor.

Hot Spot #1 – Descriptive Statistics Versus Inferential Statistics

There is an easy way to understand the difference between descriptive statistics and inferential statistics. We simply look at the root word in each case. Describe is the root of descriptive, and infer is the root of inferential.

In descriptive statistics, we are trying to describe the population. We do this by either calculating sample statistics as estimates of what we will later define as population parameters or through the use of graphs that show how the data is distributed. In inferential statistics, we draw conclusions based on the analysis of the data.

A very easy way to distinguish between a statistic and parameter is to remember that the word sample starts with an s just like statistics starts with an s. In the same way, both population and parameter start with the letter p. In fact, it might be a good idea to get into the habit of saying the phrases "sample statistic" and "population parameter" rather than the words statistics and parameter. Admittedly, this oversimplifies the problem, but a little association may go a long way to keeping the terms in perspective. On a more serious level, we need to look at the relationship between a sample statistic and a population parameter in both descriptive and inferential statistics.

In the branch of descriptive statistics, it is important to understand that we seldom know the value of a parameter. We are trying to estimate the parameter which is a measurable characteristic of the population. Since most populations are quite large, it would be difficult, if not impractical, to measure a characteristic of the population. Therefore, we use the sample statistic as an estimate of the population parameter. In the beginning, we will call this estimate a point estimate. Later on, we will extend our concept of an estimate to what we will call an interval estimate.

We have a different situation in inferential statistics where we assume a value of the population parameter. This is part of the inferential area labeled hypothesis testing. The hypothesis is that the population parameter is equal to the assumed value. The sample statistic is calculated and compared to the hypothesized value of the population parameter. If the sample statistic differs greatly from the assumed value of the population parameter, we say that we reject the hypothesis. Much of our work in the following chapters is an effort to quantify the process of hypothesis testing. As we do so, it becomes very important to distinguish between the value of the sample statistic and the value of the population parameter.

Hot Spot #2 – A Sample Versus A Population

Mann
Section 1.3

We define a population as the entire collection of all the elements we are interested in. Usually, we are interested in some measurable characteristic in the form of numerical data. When it is possible and practical, we study populations directly. The US Census is a familiar effort to collect data from an entire population. But most of us do not have the resources of the federal government available to do a census. So, we must settle for collecting smaller sets of data.

This smaller collection of data is the sample. Ideally we want as large a sample as our resources allow to insure that the sample is representative of the population. We also want to avoid any bias that might result if only certain parts of the population are selected. We accomplish this by using a probability sample where the sample is obtained by a chance process. Most of the data analysis we do later on assumes that our sample is a random sample where all samples of the same size have an equal chance of being selected.

Hot Spot #3 – $\sum x^2$ Versus $\left(\sum x\right)^2$

The Greek letter sigma, Σ, is one of the most useful and important notations that we find in statistics. The symbol $\sum$ (sigma) is used to indicate a sum. The expression $\sum x$ means that we are to add up all the data values. Actually, the expression $\sum x$ is a shortened form of the following expression:

$$\sum_{i=1}^{n} x_i = x_1 + x_2 + x_3 + \cdots + x_n$$

The shortened form $\sum x$ is much easier to write and remember. The important point to remember is that whatever follows the sigma sign is to be added together. It may help to think or say "summation of x" rather than "sigma x" when we see $\sum x$.

For example, consider the following data set:

$$2, 8, 4, 9, 5, 6$$

If we want the sum of the data values, we write the expression $\sum x$ and find the summation of x as follows:

$$\sum x = 2 + 8 + 4 + 9 + 5 + 6 = 34$$

If we want to square the sum of the data values, we use a set of parentheses to indicate that the sum is found first, and we write the expression $\left(\sum x\right)^2$. We then square the sum as shown below:

$$\left(\sum x\right)^2 = (34)^2 = 1{,}156$$

If we want to find the sum of squares, we write the expression $\sum x^2$. We first square each data value. We then add the squares to get our answer:

$$\sum x^2 = 2^2 + 8^2 + 4^2 + 9^2 + 5^2 + 6^2$$
$$= 4 + 64 + 16 + 81 + 25 + 36$$
$$= 226$$

It should be very clear that the two summation expressions

$$\left(\sum x\right)^2 \text{ and } \sum x^2$$

do not give the same value. We use both of these expressions in the alternative formula for the variance. We will also use the expressions in the chapter near the end of the book on correlation and regression. It is important to know how to use the sigma notation correctly.

Hot Spot #3 Sample Problem and Solution: Find $\sum x$ and $\sum x^2$ for the following data set: 1, 2, 3, 4, 5

Solution: $\sum x = 1 + 2 + 3 + 4 + 5 = 15$

$$\sum x^2 = 1^2 + 2^2 + 3^2 + 4^2 + 5^2 = 1 + 4 + 9 + 16 + 25 = 55$$

Sample Problem and Answer: Find $\sum x$ and $\left(\sum x\right)^2$ for the following data set: 1.1, 2.3, 1.2, 2.5, 1.0

Answer: $\sum x = 8.1$

$$\left(\sum x\right)^2 = (8.1)^2 = 65.61$$

Sample Problem and Answer: Find $\sum x$, $\sum x^2$, and $\left(\sum x\right)^2$ for the following data set: $\quad$ -2, -1, 0, 1, 2

Answer: $\quad \sum x = 0$

$\quad\quad\quad \sum x^2 = 10$

$\quad\quad\quad \left(\sum x\right)^2 = (0)^2 = 0$

CHAPTER 1 DISCUSSION QUESTIONS

These questions may be used in your study group or simply as topics for individual reflection. Whichever you do, take time to explain verbally each topic to insure your own understanding. Since these questions are intended as topics for discussion, answers to these questions are not provided. If you find that you are not comfortable with either your answers or that your group has difficulty with the topic, take time to meet with your professor to get help.

1. What is statistics?

2. What is the difference between descriptive statistics and inferential statistics?

3. What is the difference between a population and a sample?

4. What is a survey?

5. What is the difference between a sample and a census?

6. What is the difference between a simple random sample and a representative sample?

7. How does one select a simple random sample?

8. What is the difference between a discrete variable and a continuous variable?

9. What is the difference between a quantitative variable and a qualitative variable?

10. What is the difference between cross-sectional data and time series data?

$\Sigma\circ$

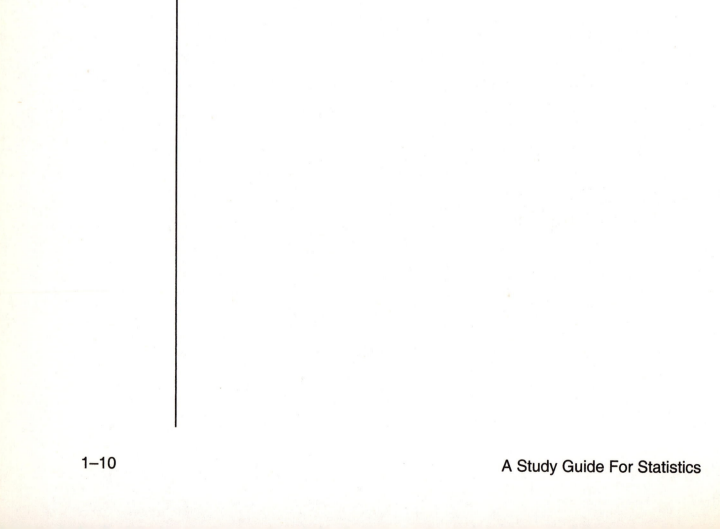

chapter two

ORGANIZING DATA

"Data! data! data!"
he cried impatiently.
"I can't make bricks without clay!"

– Sherlock Holmes

INTRODUCTION

In this chapter we begin working with data. The data may be **qualitative** or **quantitative** and come from either what is called a **population** of data values or simply a subset of the population that we call a **sample**. We will generally use the letters x or y to represent the unspecified data values. The idea of frequency will be used to both summarize and organize a collection of data values. We accomplish this by creating frequency tables and drawing graphs.

We obtain a **frequency table** by listing the possible data values together with the frequency, or number of times the data value is found in the collection of data. The result is called a **frequency distribution**. We will focus on the idea of the frequency distribution throughout the remaining chapters of the book.

There is a second type of frequency distribution we create by first dividing the collection of data into intervals called **classes**. We then find the number of data values in each class. A large collection of data can be reduced to a smaller number of classes to get a better idea of how the data is distributed. We call the result a **grouped frequency distribution** where **class limits** and **class boundaries** are used to create the classes. Each class has the same **class width,** and all the data in each of the classes is represented by the midpoint, the **class mark**, of the interval.

We use several types of graphs to summarize and organize data collections. You may already be familiar with some of the graphs that are used with qualitative data , such as the **pie chart**, pictogram, broken line graph, and the **bar graph**. When we work with quantitative data, the broken line graph and bar graph become the frequency polygon and frequency histogram.

The **histogram** involves drawing a bar graph where the bars are contiguous (touching) and represent the **frequency, relative frequency, cumulative frequency,** or **cumulative relative frequency** of the data in each class of the grouped data frequency distribution. If we use a dot at the mid point of each class rather than a bar to indicate the frequency of each interval, we can join the dots to create a broken line graph called the **frequency polygon or relative frequency polygon**.

Both the histogram and the polygon are used to suggest the shape of the distribution of our data values. We are interested in finding out if the distribution is a symmetric bell-shaped curve or if there is more data on one of the sides of the distribution. We refer to this second situation as a **skewed distribution**. Most of the results we obtain in later chapters require that we have a bell shaped distribution.

In this chapter we will also work with a simple way of examining the shape of a distribution using what is called a **stem-and-leaf display** from the recently developed field of EDA (exploratory data analysis). We can create the stem-and-leaf display as we record data. The resulting display resembles a histogram that we can use to determine if the shape of the distribution looks like a symmetric bell shaped curve.

CHAPTER 2 HOT SPOTS

1. **Number Of Intervals In A Grouped Frequency Table.**
   Starts on **page 2–4.** Problems on **pages 2–5, 2–6.** Mann 2.3.2

2. **Class Boundaries Versus Class Limits.**
   Starts on **page 2–6.** Problems on **pages 2–9, 2–10, 2–11.**
   Mann 2.3.1

3. **Frequency Versus Relative Frequency.**
   Starts on **page 2–11.** Problems on **pages 2–13, 2–15.**
   Mann 2.2.2, 2.3.3, 2.3.4

4. **Class Marks Versus Class Boundaries.**
   Starts on **page 2–16.** Problems on **pages 2–17, 2–18.** Mann 2.3.1

5. **Finding The Class Width.**
   Starts on **page 2–19.** Problems on **pages 2–19, 2–20.**
   Mann 2.3.1, 2.3.2

6. **Labeling The Horizontal Axis For Graphs.**
   Starts on **page 2–21.** Problems on **pages 2–24, 2–25, 2–26.**
   Mann 2.3.4

7. **Number Of Degrees In Each Sector Of A Pie Chart.**
   Starts on **page 2–28.** Problems on **pages 2–29, 2–30, 2–31.**
   Mann 2.2.3

If you find other HOT SPOTS, write them down and use them as a focus of your discussions in the study group. Or you can use the HOT SPOT as the topic for a help session with your professor.

Hot Spot #1 – Number Of Intervals In A Grouped Frequency Table

We often have a large collection or set of data. It is then difficult to see how the data is distributed. We try to simplify the data by dividing a large collection of data into a smaller number of intervals. The argument is that it is much easier to examine the shape of a distribution when working with between 5 and 15 intervals than with a larger number of data. Of course, this brings up the issue of what is meant by a large number of data. If there are less than 30 pieces of data, there really is not much of a need to regroup data. Clearly, there is a need to regroup the data if there are more than 100 pieces of data. It can be confusing for the beginning statistics student when trying to choose the appropriate number of intervals for larger data sets.

There is no absolute rule that needs to be followed in selecting the number of intervals. However, there are several different guidelines that can be followed. One of the simplest rules is to take the square root of the sample size to find the number of intervals to use. This rule works fairly well for data sets that are larger than 30 and smaller than 150. For example if the collection had 50 pieces of data in it, we would use 7 intervals.

$$\text{Number of intervals} = \sqrt{50} \approx 7$$

Unfortunately this rule tends to give more intervals than is really necessary when we work with very large data sets.

A more effective rule uses the powers of 2 to find the number of intervals. We first need to make a reference list of the powers of 2:

The First Ten Powers of 2	
$2^1 = 2$	$2^6 = 64$
$2^2 = 4$	$2^7 = 128$
$2^3 = 8$	$2^8 = 256$
$2^4 = 16$	$2^9 = 512$
$2^5 = 32$	$2^{10} = 1{,}024$

Table 2–1:
The First Ten
Powers of 2

We really do not need the first column because we do not worry about regrouping small collections of data. The second column would be used as shown in the sample problem that follows. Briefly stated we find the closest power of 2 that matches the number of pieces of data in our collection. The exponent becomes the number of intervals we use in our grouped frequency distribution. It should be noted that the text does not use this rule, and the number of classes in the exercises may not agree with this approach. Nevertheless, the powers of 2 approach is easy to use when we are asked to determine the number of intervals or classes.

Hot Spot #1 Sample Problem and Solution: Determine the number of intervals for a data collection with 200 pieces of data.

Solution: Step 1.   200 is between 128 and 256.

Step 2.   200 is closer to 256.

Step 3. $256 = 2^8$.

Step 4.   We use 8 intervals.

Sample Problem and Answer: How many intervals should be used for a data collection containing 75 pieces of data?

Answer: 75 is closest to $2^6 = 64$. We use 6 intervals.

Sample Problem and Answer: The data collection contains 500 pieces of data. How many intervals should be used?

Answer: 500 is closest to $2^9 = 512$. We use 9 intervals.

✱

*Mann
Section 2.3.1*

Hot Spot #2 – Class Boundaries Versus Class Limits

The difference between a class boundary and class limit often causes trouble when we first construct a grouped frequency distribution table. An easy distinction can be made by looking at any two consecutive intervals. If the intervals share the same value, the numbers are boundaries.

As an example, we will consider a grouped frequency distribution table of data taken from a collection of highway speeds. When we use class boundaries, any two adjacent intervals or classes contain the same value. The lower and upper values of each interval are called the lower and upper class boundaries. When we use class limits to construct the same grouped frequency distribution table, there is a break between the limits going from one interval to the next.

The following grouped frequency distribution table illustrates the difference between class boundaries and class limits.

	Highway Speeds		
Class	Class Boundaries	Class Limits	Frequency
1	24.5–34.5	25–34	7
2	34.5–44.5	35–44	13
3	44.5–54.5	45–54	25
4	54.5–64.5	55–64	50
5	64.5–74.5	65–74	5

Table 2–2: Sample Grouped Frequency Distribution Table

There are a few advantages to using class limits. The first one is that a data value is clearly in a particular interval or class. A second point is that a person using the table to create a grouped frequency distribution might find it easier to determine which interval each piece of data is in when class limits are used.

The disadvantages of using class limits involves both the graphs we do in this chapter and the data analysis that will be done in the next chapter. In each case, we will need to change the class limits to class boundaries. Regardless of the advantages or disadvantages of class limits, both class limits and class boundaries are used in statistics. It is important to be able to go from one form to the other. It is easy to create the class boundaries from class limits. All we need to do is find the value halfway between the upper class limit of one class and the lower class limit of the next class. For example in the intervals used for highway speeds in the above example, 34 is the upper class limit of the first interval, and 35 is the lower class limit of the next interval.

Class Limits
25–34
35–44

The upper class boundary of the first class and the lower class boundary of the second interval becomes 34.5; it is the value halfway between 34 and 35. The same thing is done to get the upper class boundary of the second interval. We look at the next interval.

45–54

The class boundary 44.5 is halfway between 44 and 45. We continue in the same way until each interval is written with class boundaries. Our only remaining problem is with the first and last intervals. In order to be consistent and have each interval the same width, the lower class boundary of the first interval is started at 24.5; the first interval now goes from 24.5 to 34.5. Likewise, the upper class boundary of the last interval is written as 74.5; the last interval now goes from 64.5 to 74.5.

There is not a practical need to change from class boundaries to class limits. If we decided to change the boundaries back to limits, all we would need to do is add and subtract the 0.5 from the boundaries.

Of course this brings up the question of what is done if the original limits had been given in tenths rather than units. For example, if the highway speeds had been recorded to the nearest tenth, the original limits might have been given as follows:

24.6–34.5
34.6–44.5
44.6–54.5
etc.

Hot Spot #2 Sample Problem and Solution: Given the table:

Highway Speeds	
Class	Class Limits
1	24.6–34.5
2	34.6–44.5
3	44.6–54.5
4	54.6–64.5
5	64.6–74.5

Table 2–3a: Highway Speeds w/Class Limits

change the class limits to class boundaries.

Solution: We could change to class boundaries by again finding a value halfway between class limits.

Step 1. Using the first and second intervals, we could find the value halfway between 34.5 and 34.6. The class boundary is 34.55; a short cut for finding "halfway" is

$$\frac{34.5 + 34.6}{2} = 34.55.$$

Step 2. Using the second and third intervals, the class boundary 44.55 is the value halfway between 44.5 and 44.6.

Step 3. We continue in the same way finding the class boundaries 54.55 and 64.55.

Step 4. We change the lower class limit of the first interval to 24.55 and the upper class limit of the last interval to 74.55 to maintain equal widths for each interval.

Step 5. The table now reads as follows:

Highway Speeds	
Class	Class Boundaries
1	24.55–34.55
2	34.55–44.55
3	44.55–54.55
4	54.55–64.55
5	64.55–74.55

Table 2–3b: Highway Speeds w/Class Boundaries

Sample Problem and Answer: Given the following table

Highway Speeds	
Class	Class Boundaries
1	69.5–79.5
2	79.5–89.5
3	89.5–99.5

Table 2–4a

change the class boundaries to class limits.

Answer:

Highway Speeds	
Class	Class Limits
1	70–79
2	80–89
3	90–99

Table 2–4b

Sample Problem and Answer: Given the table:

City Speeds

Class	Class Limits
1	15.4–18.7
2	18.8–22.1
3	22.2–25.5

Table 2–5a

change the class limits to class boundaries.

Answer:

City Speeds

Class	Class Boundaries
1	15.35–18.75
2	18.75–22.15
3	22.15–25.55

Table 2–5b

Hot Spot #3 – Frequency Versus Relative Frequency

When we count the number of times we find a particular value in a collection of data, we are getting what is called the frequency of that piece of data. We also find the frequency for each interval in a grouped frequency table by counting how many pieces of data lie in the interval. The relative frequency of either a piece of data or an interval is simply the frequency divided by the number of data in the sample. If we let n equal the sample size, then the formula for relative frequency is as follows:

$$\text{relative frequency} = \frac{f}{n}$$

✳
Mann
Sections 2.2.2,
2.3.3, And 2.3.4

Organizing Data

2–11

The idea of relative frequency is necessary to create some of the graphs that we use in this chapter. But more importantly, the concept of relative frequency plays an even greater role in what we will call experimental probability in a later chapter. Therefore, relative frequency is an important idea that we need to understand.

If we are given a sample (collection of data) like the following

$$3, 5, 2, 3, 4, 5, 4, 5, 7, 5, 2, 5, 3, 6, 4, 4, 4, 5, 5, 3$$

we can make a frequency table by listing each possible data value and the frequency for each data value as follows:

Data Value	Frequency
2	2
3	4
4	5
5	7
6	1
7	1

Table 2–6a: Values and Frequencies

We find the relative frequency for each data value by dividing each frequency by $n=20$, the sample size. Note: Be sure to first check that $\Sigma f = 20$. An easy way to change each frequency to relative frequency is to create a fraction that has the 20 in the denominator as follows:

Data Value	Frequency	Relative Frequency
2	2	$\frac{2}{20}$
3	4	$\frac{4}{20}$
4	5	$\frac{5}{20}$
5	7	$\frac{7}{20}$
6	1	$\frac{1}{20}$
7	1	$\frac{1}{20}$

Table 2–6b: Relative Frequencies Added

If we want, we can change each of the fractions to a decimal. When we add all of the relative frequency values together, we should get a sum of 1. Since the sum of the relative frequencies is 1, we can check our work each time by adding the relative frequencies for all of the data values or intervals:

Relative Frequency
$\frac{2}{20} = 0.10$
$\frac{4}{20} = 0.20$
$\frac{5}{20} = 0.25$
$\frac{7}{20} = 0.35$
$\frac{1}{20} = 0.05$
$\frac{1}{20} = 0.05$

sum = 1.00 ✔

Table 2–6c: Relative Frequencies Checked

It is very important when drawing either a polygon or histogram to check if we are to use frequencies or relative frequencies. When we use frequencies, we get frequency polygons and frequency histograms. In each case the vertical axis gives the frequency for each interval in the grouped frequency table. If the relative frequency is used, the vertical axis gives the relative frequency; the graphs are called relative frequency polygons and relative frequency histograms.

Hot Spot #3 Sample Problem and Solution: Find the relative frequency for each class in Table 2-7a and check that the sum of the relative frequencies totals 1.

City Speeds		
Class	Class Boundaries	Frequency
1	20.5–24.5	3
2	24.5–28.5	5
3	28.5–32.5	9
4	32.5–36.5	8

Table 2–7a: City Speeds w/Frequencies

Solution: Step 1.   The sample size is 25

Step 2.   Change each frequency to a fraction with the denominator of 25 and write the decimal value

$$\frac{3}{25} = 0.12$$

$$\frac{5}{25} = 0.20$$

$$\frac{9}{25} = 0.36$$

$$\frac{8}{25} = 0.32$$

Step 3.   Check the sum of the values for equality to 1.0

$$\text{sum} = 1.00 \checkmark$$

Resulting Table:

City Speeds		
Class	Class Boundaries	Relative Frequency
1	20.5–24.5	$\frac{3}{25}$
2	24.5–28.5	$\frac{5}{25}$
3	28.5–32.5	$\frac{9}{25}$
4	32.5–36.5	$\frac{8}{25}$

Table 2–7b: City Speeds w/Relative Frequencies

Sample Problem and Answer: Find both the frequency and relative frequency for each data value in the following sample:

$$5, 3, 5, 4, 7, 5, 4, 6, 5, 6$$

Answer:

Data Value	Frequency	Relative Frequency
3	1	$\frac{1}{10} = 0.1$
4	2	$\frac{2}{10} = 0.2$
5	4	$\frac{4}{10} = 0.4$
6	2	$\frac{2}{10} = 0.2$
7	1	$\frac{1}{10} = 0.1$

Table 2–8

$$\text{sum} = 1.0 \checkmark$$

Sample Problem and Answer: Find both the frequency and relative frequency for each class in the following sample:

$$1.3, 2.6, 1.5, 3.3, 3.8, 1.7, 2.1, 3.2, 4.9, 1.1$$

Class	Class Boundaries	Frequency	Relative Frequency
1	1.05–2.05		
2	2.05–3.05		
3	3.05–4.05		
4	4.05–5.05		

Table 2–9a

Answer:

Class	Class Boundaries	Frequency	Relative Frequency
1	1.05–2.05	4	$\frac{4}{10} = 0.4$
2	2.05–3.05	2	$\frac{2}{10} = 0.2$
3	3.05–4.05	3	$\frac{3}{10} = 0.3$
4	4.05–5.05	1	$\frac{1}{10} = 0.1$

sum = 1.0 ✔

Table 2–9b

✳
*Mann
Section 2.3.1*

Hot Spot #4 – Class Marks Versus Class Boundaries

We divide large data sets into a smaller number of intervals or classes to help us interpret the distribution. We no longer have the individual pieces of data when we create a grouped frequency distribution. In order to represent the data in each class we use what is called a class mark. The class mark is the number halfway between the lower and upper class limits of each class. We can also find the class mark by using the number halfway between the lower and upper class boundaries of each class. A helpful point in finding class marks is to remember that the difference between any two consecutive class marks must be the same as the width of each class or interval. We treat all the data in the class as though each piece of data had the value of the class mark.

It is also important to remember when we use class marks in the graphs we construct. We use class marks on the horizontal axis when we are creating polygons. We use class boundaries on the horizontal axis when we are creating histograms.

Hot Spot #4 Sample Problem and Solution: Find the class marks for the following grouped frequency distribution table:

Highway Speeds		
Class	Class Limits	Class Mark
1	50–54	
2	55–59	
3	60–64	
4	65–69	

Table 2–10a: Finding Class Marks

Solution: Step 1. The value halfway between 50 and 54 is 52. A short cut is to use $\dfrac{50 + 54}{2} = 52$.

Step 2. The width of each class is 55-50 = 5, so the next class mark is 52+5 = 57. We can check to see that 57 is halfway between 55 and 59.

Step 3. The remaining class marks are 57+5 = 62 and 62+5 = 67.

Step 4. We now fill in the class marks in the table:

Highway Speeds		
Class	Class Limits	Class Mark
1	50–54	52
2	55–59	57
3	60–64	62
4	65–69	67

Table 2–10b: Class Marks Found

Sample Problem and Answer: Find the class marks for the following grouped frequency table:

Class	Class Boundaries	Class Mark
1	10.5–20.5	
2	20.5–30.5	
3	30.5–40.5	
4	40.5–50.5	

Table 2–11a

Answer:

Class	Class Boundaries	Class Mark
1	10.5–20.5	15.5
2	20.5–30.5	25.5
3	30.5–40.5	35.5
4	40.5–50.5	45.5

Table 2–11b

Sample Problem and Answer: Find the class marks for the following grouped frequency distribution table:

Class	Class Limits	Class Mark
1	25–28	
2	29–32	
3	33–36	

Table 2–12a

Answer:

Class	Class Limits	Class Mark
1	25–28	26.5
2	29–32	30.5
3	33–36	34.5

Table 2–12b

Hot Spot #5 – Finding The Class Width

✳
*Mann
Sections
2.3.1 And 2.3.2*

Finding the class width can be very simple. We find the class width by taking the difference of either two consecutive upper class limits or two consecutive lower class limits. If the grouped frequency table uses class boundaries, we simply find the difference of either two consecutive upper class boundaries or two consecutive lower class boundaries. In fact, it can even be easier to find the class width for a table with class boundaries; the difference between the upper and lower boundary of any class will give us the class width. We must be careful, however, as this rule will not work when our table uses class limits.

Knowing the class width helps us check for errors when we construct a grouped frequency table. We can do quick visual checks to make sure the difference between any two consecutive upper limits or boundaries is the same as the class width. The same should be true for the difference of any two consecutive lower class limits or class boundaries. We also should have the difference of any two consecutive class marks the same as the class width.

Hot Spot #5 Sample Problem and Solution: Find the class width for the following grouped frequency table:

Class	Class Limits	Frequency	Class Mark
1	1.25–1.49	2	1.37
2	1.50–1.74	7	1.62
3	1.75–1.99	3	1.87

*Table 2–13:
Finding Class
Width Using
Class Limits*

Solution: Step 1. The difference between the first two lower limits is 1.50-1.25 = 0.25.

Step 2. The class width is 0.25.

Step 3. A visual check of the difference of all consecutive limits and class marks shows each difference is 0.25.

Sample Problem and Answer: Find the class width for the following grouped frequency distribution table:

Class	Class Boundaries	Frequency	Class Mark
1	3.25–3.35	1	3.3
2	3.35–3.45	3	3.4
3	3.45–3.55	2	3.5

Table 2–14

Answer: The class width = 0.1 and the visual check of the difference between consecutive boundaries and class marks shows that all the differences are 0.1.

Sample Problem and Answer: Find the class width for the following grouped frequency distribution table:

Class	Class Limits	Frequency	Class Mark
1	1.0–1.9	3	1.45
2	2.0–2.9	2	2.45
3	3.0–3.9	8	3.45
4	4.0–4.9	5	4.45

Table 2–15

A Study Guide For Statistics

Answer: The class width equals the difference of 2.0-1.0 = 1. A visual
check of the difference between all consecutive limits and class
marks shows that all of the differences are 1.0.

Hot Spot #6 – Labeling The Horizontal Axis For Graphs

✷
Mann
Section 2.3.4

There are several different types of graphs we create in this
chapter. The polygon and the histogram are particularly important
graphs because they help us decide whether or not the distribution is
bell shaped. Both the polygon and the histogram can be drawn in
several versions; namely as frequency, cumulative frequency, relative
frequency, or cumulative relative frequency polygons and histograms.
We will focus on the frequency and relative frequency graphs because
they give us a better idea of what the distribution of data looks like.

In the frequency and relative frequency polygon and histogram,
the vertical axis is either the frequency or relative frequency. However,
there is a difference in the choice of the horizontal axis for the two
graphs. The following grouped frequency distribution table will be used
in the discussion that follows:

Autobahn Speeds in Germany			
Class	Class Boundaries	Frequency	Class Mark
1	45.5-55.5	3	50.5
2	55.5-65.5	7	60.5
3	65.5-75.5	12	70.5
4	75.5-85.5	6	80.5
5	85.5-95.5	2	90.5

Table 2–16: Grouped Frequency Distribution Table Of Autobahn Speeds in Germany

In the histogram, the bar represents the width of the class or interval. As such, the class boundaries are used on the horizontal axis of the histogram. The frequency histogram for the grouped frequency distribution table above would look like the following graph:

Frequency Histogram

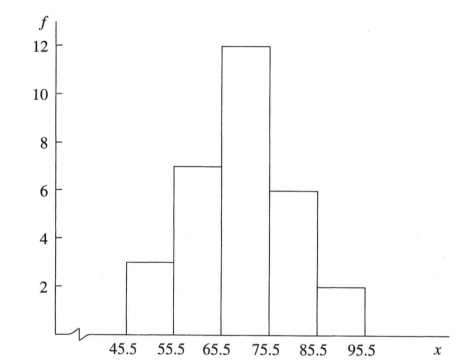

Figure 2–1: Autobahn Speeds Frequency Histogram

Autobahn Speeds in Germany

In the polygon, a point is used to represent the frequency or relative frequency of the entire interval. As such, the point is placed above the class mark that represents all of the data in the class. The frequency polygon for the above grouped frequency distribution table (Table 2–16) would like the following graph:

Frequency Polygon

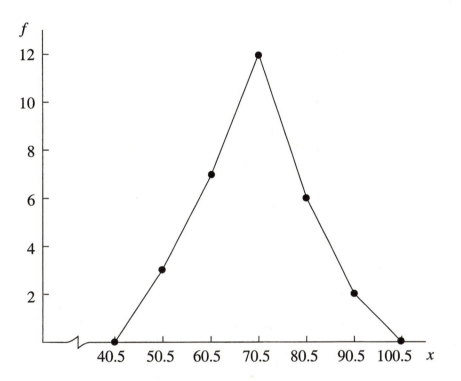

Autobahn Speeds in Germany

Figure 2–2:
Autobahn Speeds
Frequency
Polygon

It is easy to adjust either of the above graphs to make them either a relative frequency histogram or a relative frequency polygon. All that has to be done is to change each value on the vertical axis to a fraction by rewriting it as a number over the sample size, which is 30 in this example.

Hot Spot #6 Sample Problem and Solution: Given the above grouped frequency distribution table (Table 2–16), draw the relative frequency polygon.

Solution:

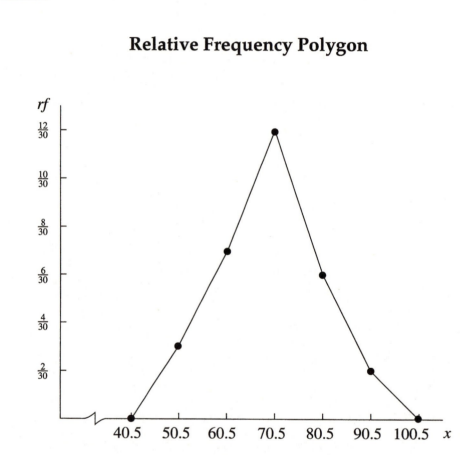

Figure 2–3:
Autobahn Speeds
Relative
Frequency
Polygon

A Study Guide For Statistics

Sample Problem and Answer: Given the above grouped frequency
distribution table (Table 2–16), draw the relative frequency histogram.

Answer:

Relative Frequency Histogram

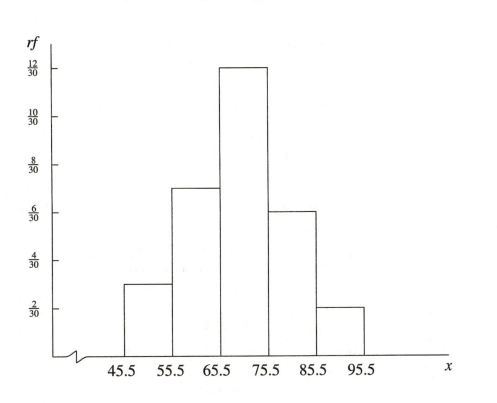

Autobahn Speeds in Germany

Figure 2–4:
Autobahn Speeds
Relative
Frequency
Histogram

Sample Problem and Answer: Change the following frequency polygon to a relative frequency polygon:

Frequency Polygon

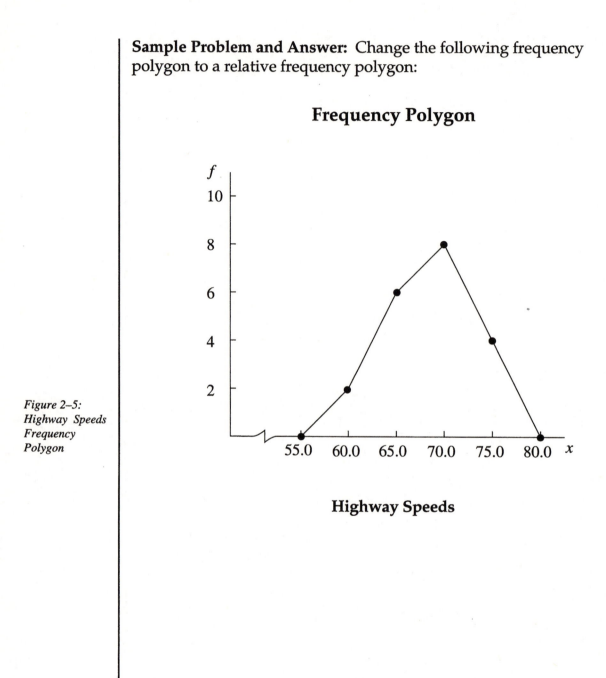

Figure 2–5:
Highway Speeds
Frequency
Polygon

Highway Speeds

Answer: Divide each frequency by 20.

Relative Frequency Polygon

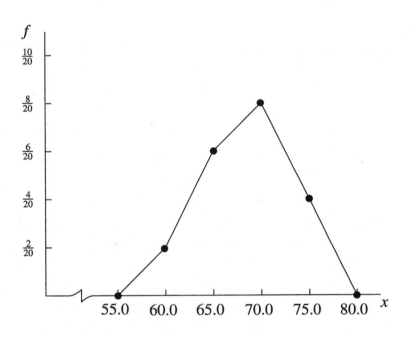

Highway Speeds

Figure 2–6:
Highway Speeds
Relative
Frequency
Polygon

Hot Spot #7 – Number Of Degrees In Each Sector Of A Pie Chart

We use pie charts when we are working with data that belongs to non-numerical categories. The pie chart is used to show the percentage of the sample in each category.

The process involves the following steps:

Step 1.   Draw a circle using a compass or protractor.

Step 2.   Find the relative frequency for each category.

Step 3.   Multiply each relative frequency by 360, the number of degrees in a circle.

Step 4.   Use a protractor to construct a sector for each category; the size of the angle for each sector is given by the answers in the preceding step.

Step 5.   Label each sector with the appropriate category and indicate the relative frequency as a percent.

The next sample problem illustrates how we should use the five steps to create a pie chart. One final thought before we illustrate the process: be sure to title the pie chart so that the reader understands what is being represented. The actual size of the finished pie chart is a matter of personal taste. However, pie charts are often done in color so that they fill an $8\frac{1}{2}$"×11" format that is made into an acetate or slide that is used in a formal presentation, such as in a business meeting. Most of the pie charts used in the business community are created using a computer running a spreadsheet program, such as Lotus® 1-2-3®, or Microsoft® Excel.

Lotus and 1-2-3 are registered trademarks of Lotus Development Corporation.
Microsoft is a registered trademark of Microsoft Corporation.

Hot Spot #7 Sample Problem and Solution: Given the following information about education costs at a university for the year, create a pie chart.

Housing costs	$6,000.00
Food costs	$3,000.00
Car expenses	$1,000.00
Tuition/Books	$2,000.00

Solution: Step 1. The circle is drawn below in step 4.

Step 2. The relative frequencies are determined.

Category	Relative Frequency
Housing Costs	0.500
Food Costs	0.250
Car Expenses	0.083
Tuition/Books	0.167

Step 3. Each relative frequency is multiplied by 360°, and rounded to the nearest degree.

Category	Relative Frequency
Housing Costs	$0.500 \times 360° \cong 180°$
Food Costs	$0.250 \times 360° \cong 90°$
Car Expenses	$0.083 \times 360° \cong 30°$
Tuition/Books	$0.167 \times 360° \cong 60°$

Step 4. Each sector is shown in the following circle with the degrees indicated. You should use a protractor to verify the measurements.

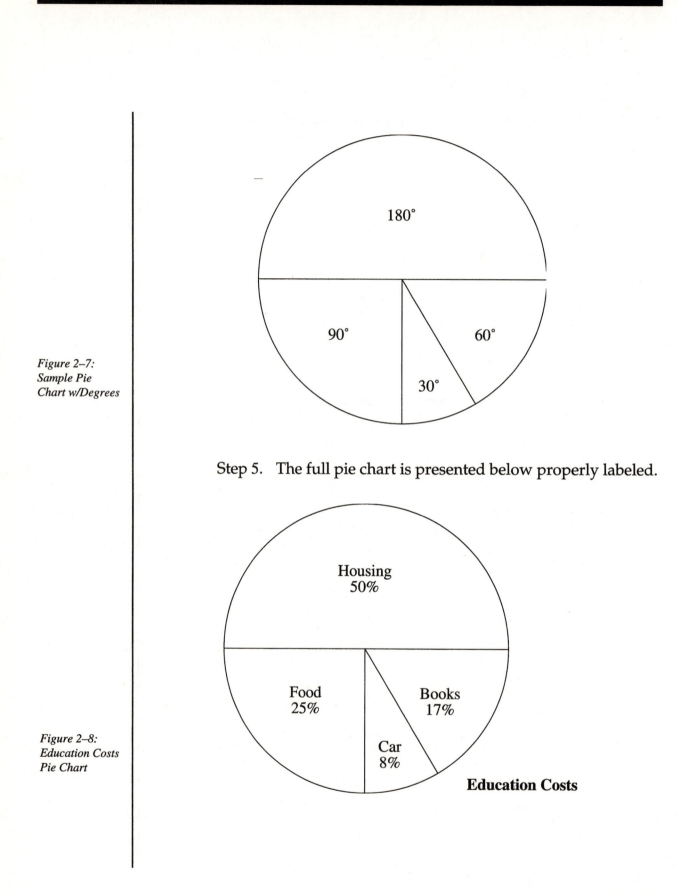

Figure 2–7:
Sample Pie
Chart w/Degrees

Step 5.   The full pie chart is presented below properly labeled.

Figure 2–8:
Education Costs
Pie Chart

Sample Problem and Answer: Draw a pie chart for the following grade distribution:

A's	5
B's	10
C's	30
D's	5

Answer: Step 1. The circle is drawn using a compass.

Step 2. The relative frequencies are obtained.

A's	0.10
B's	0.20
C's	0.60
D's	0.10

Step 3. The degrees in each sector are calculated.

A's	$0.10 \times 360° \cong 36°$
B's	$0.20 \times 360° \cong 72°$
C's	$0.60 \times 360° \cong 216°$
D's	$0.10 \times 360° \cong 36°$

Step 4. The sectors are drawn in the circle using a protractor.

Step 5.   The labels are then added to the chart.

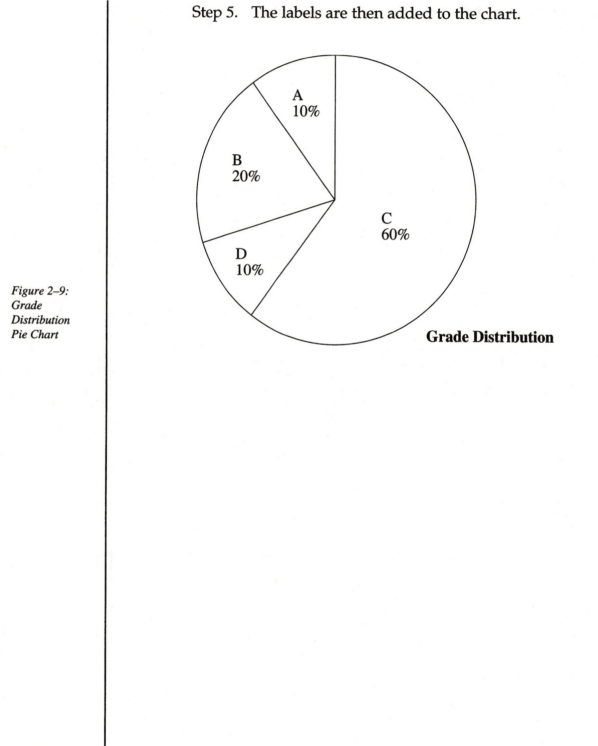

Figure 2–9:
Grade
Distribution
Pie Chart

A Study Guide For Statistics

CHAPTER 2 DISCUSSION QUESTIONS

These questions may be used in your study group or simply as topics for individual reflection. Whichever you do, take time to explain verbally each topic to insure your own understanding. Since these questions are intended as topics for discussion, answers to these questions are not provided. If you find that you are not comfortable with either your answers or that your group has difficulty with the topic, take time to meet with your professor to get help.

1. What is a frequency distribution for qualitative data?

2. What is the difference between a frequency distribution and a grouped frequency distribution for quantitative data?

3. What is the difference between a class limit and a class boundary?

4. How do you determine the number of intervals or classes to use with a grouped frequency distribution?

5. How do you determine the class width of a grouped frequency distribution?

6. What is a cumulative frequency distribution?

7. What is the difference between frequency and relative frequency?

8. Why is it important to examine the shape of the frequency polygon?

9. What is the difference between a stem and a leaf?

10. How can a truncated bar graph be used to misrepresent the data?

CHAPTER 2 TEST

1. In constructing a grouped frequency distribution, how many intervals or classes should be used for 100 pieces data?

2. What is the class mark for the class with the lower class limit of 35 and the upper class limit of 43?

3. If the sample size is 50, what is the relative frequency of the class with a frequency of 12?

Use the grouped frequency distribution table below for problems 4 to 8:

Class	Class Boundaries	Frequency	Class Mark	Relative Frequency
1	14.5–19.5	2	17	0.2
2	19.5–24.5	7	22	0.7
3	24.5–29.5	1	27	0.1

4. Rewrite the second class boundary using class limits.

5. Find the class width.

6. Find the cumulative frequency of the second class.

7. Draw the frequency polygon.

8. Draw the relative frequency histogram.

9. Use the following data to construct a stem-and-leaf display:

    31 25 34 22 34 46 52 33 49 28 50 37 41 18 62

10. Draw a pie chart for the following data indicating majors:

    English 20   Science 5   Business 15   Art/Music 10

CHAPTER 2 TEST Questions and Answers

1. In constructing a grouped frequency distribution, how many intervals or classes should be used for 100 pieces data?

Answer: 100 is between $2^6 = 64$ and $2^7 = 128$. It is closest to 128, so we use 7 classes.

2. What is the class mark for the class with the lower class limit of 35 and the upper class limit of 43?

Answer: The class mark is 39. It is halfway between 35 and 43.

3. If the sample size is 50, what is the relative frequency of the class with a frequency of 12?

Answer: The relative frequency is $\dfrac{12}{50} = 0.24$

Use the grouped frequency distribution table below for problems 4 to 8:

Class	Class Boundaries	Frequency	Class Mark	Relative Frequency
1	14.5–19.5	2	17	0.2
2	19.5–24.5	7	22	0.7
3	24.5–29.5	1	27	0.1

4. Rewrite the second class using class limits.

Answer: 20-24

5.   Find the class width.

Answer: The class width is 5.

6.   Find the cumulative frequency of the second class.

Answer: 9

7.   Draw the frequency polygon.

Answer:

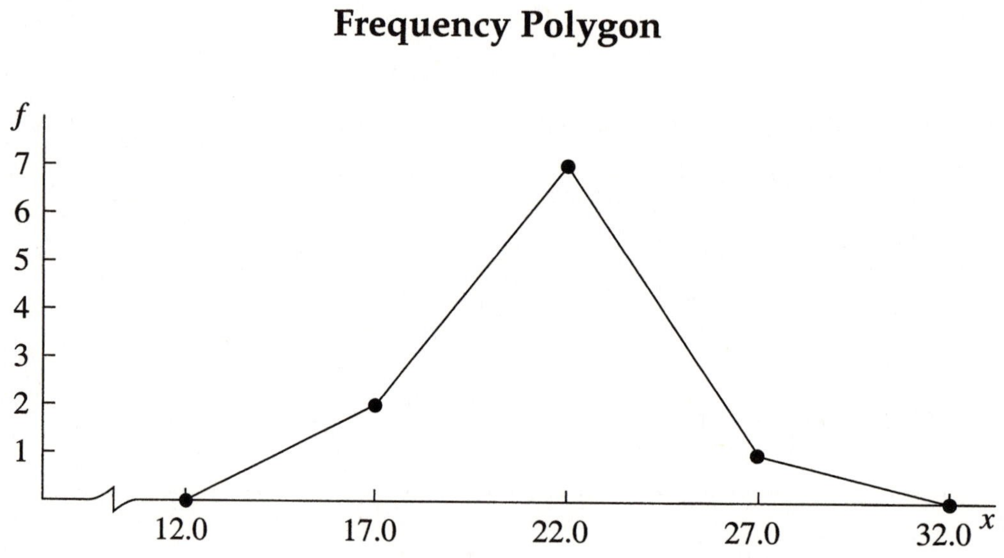

Frequency Polygon

Figure 2–10: Test Answer Frequency Polygon

8.   Draw the relative frequency histogram.

Answer:

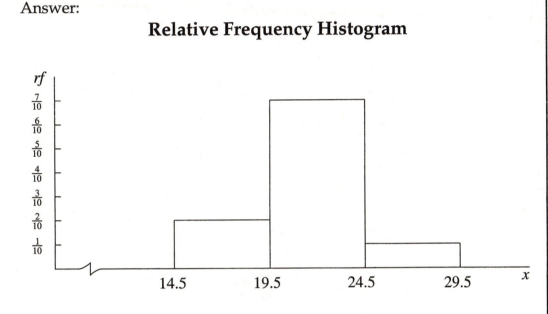

Relative Frequency Histogram

Figure 2–11:
Test Answer
Relative
Frequency
Histogram

9.   Use the following data to construct a stem-and-leaf display:

31 25 34 22 34 46 52 33 49 28 50 37 41 18 62

Answer:

Stems	Leaves
1	8
2	5, 2, 8
3	1, 4, 4, 3, 7
4	6, 9, 1
5	2, 0
6	2

Stem-and-leaf display

Figure 2–12:
Test Answer
Stem-and-leaf
display

10. Draw a pie chart for the following data indicating majors:

English 20   Science 5   Business 15   Art/Music 10

Answer:

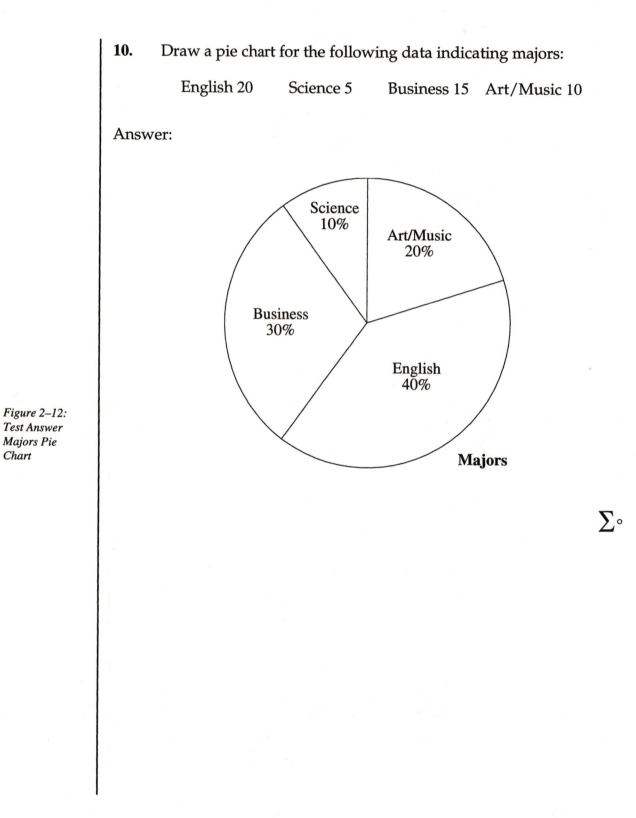

Figure 2–12: Test Answer Majors Pie Chart

$\Sigma\circ$

A Study Guide For Statistics

chapter three

NUMERICAL DESCRIPTIVE MEASURES

There are three kinds of lies -
lies, damned lies, and statistics.

– Disraeli

INTRODUCTION

In this chapter we begin our descriptive analysis of data. We focus on measures of central tendency and measures of dispersion. We will use the Greek letters μ and σ for the population values and the English letters $\bar{x}$ and s for the sample values.

The mean, median, and mode are three different measures of central tendency that are used in this chapter. The **mean** is the usual arithmetic average that is familiar to us; the **median** is the middle value in a ranked data list; and the **mode** is the piece of data occurring most often.

The range, variance, and standard deviation are the three measures of dispersion that are used. The **range** is the highest data value minus the lowest data value; the **variance** is the average, or modified average, of the squares of the deviations of the data values from the mean; and the **standard deviation** is the square root of the variance. We find variance and standard deviation for both samples and populations.

The mean and standard deviation are very important to the work we will do in later chapters. We see the beginnings of this in the current chapter as we work with what is called the **Empirical Rule**; which applies to bell shaped frequency distributions. It tells us how often we can expect to find data in an interval about the mean. This ability to quantify how often data should be found in an interval is at the heart of inferential statistics.

We also get our first opportunity to see how we can adjust our analysis if the distribution of our data does not meet our assumption of a symmetrical bell shaped distribution from a **Normal Population**. Chebyshev's Theorem gives us a way of finding out how often we can expect to find data in an interval about the mean for any shaped distribution.

Two measures of position are presented in the chapter. They are the **percentile** and **quartile** values that we often see when we work with reports of test scores.

We also extend our work with graphs in this chapter with what is called the box-and-whisker plot ("**boxplot**"). The boxplot is an extension of what was done in the last chapter using stem-and-leaf diagrams. The boxplot provides a graphic display of the numerical features of a collection of data. The median is located at the center of the box, and the sides of the box are drawn at the first and third quartiles, called the **hinges**. **Whiskers** are drawn on each side of the box to the lowest and highest data values.

This chapter is filled with new notation, many formulas, and new terms. It is very important to become familiar with the new notation, know the terms, and learn how to use the formulas. We will continue to use what is developed in this chapter throughout the book.

CHAPTER 3 HOT SPOTS

1. **Value Of Median Versus Location Of Median.**
   Starts on **page 3–5.** Problems on **pages 3–6, 3–7.** Mann 3.1.2

2. **Effect Of Skewing On Mean and Median.**
   Starts on **page 3–8.** Mann 3.1.4

3. $\sum x^2$ **Versus** $\left(\sum x\right)^2$.
   Starts on **page 3–11.** Problems on **pages 3–12, 3–13.** Mann 3.2.2

4. $\sum f$ **In The Formulas For** $\bar{x}$ **And** s **For Grouped Data..**
   Starts on **page 3–14.** Mann 3.3.1, 3.3.2

5. **Checking To See That We Have** s **Rather Than** s^2.
   Starts on **page 3–15.** Problems on **pages 3–16, 3–17.** Mann 3.2.2

6. **The Meaning Of** $\bar{x} \pm ks$.
   Starts on **page 3–17.** Problems on **page 3–19.** Mann 3.4.1, 3.4.2

7. **Understanding Chebyshev's Theorem.**
   Starts on **page 3–20.** Problems on **page 3–23.** Mann 3.4.1

8. **When** P_k **Does Not Result In** $k\%$ **Of The Data Below** P_k.
   Starts on **page 3–24.** Mann 3.5.1, 3.5.2

9. **Rounding Off.**
   Starts on **page 3–37.** Mann 3.3

If you find other HOT SPOTS, write them down and use them as a focus of your discussions in the study group. Or you can use the HOT SPOT as a the topic for a help session with your professor.

USE THIS PAGE FOR KEEPING TRACK OF TERMS AND NOTATION

Hot Spot #1 – Value Of Median Versus Location Of Median.

✳
Mann
Section 3.1.2

The median is one of three measures of central tendency that we find in this chapter, and it is easy to find. All we have to do is rank the data and find the middle data value. The formula for the location of the median is also easy to use and gives us the location of the median as

$$\frac{n+1}{2}.$$

When the number of pieces of data is odd, the middle data value is one of the pieces of data. The only real difficulty occurs when the number of data values is even. We then find a value halfway between the two data values in the middle. The two ranked data sets that follow illustrate the difference:

Data Set 1:   3, 3, 4, 7, 9

The location of the median is $\dfrac{5+1}{2} = 3$.

The median is 4.

Data Set 2:   3, 5, 7, 9

The location of the median is $\dfrac{4+1}{2} = 2.5$.

The median is the value halfway between 5 and 7.

The median is 6.

Although finding the location of the median and the value of the median is easy to do, an error is often made in confusing the location of the median with the value of the median.

One of the possible explanations for the error is the formula used for the position of the median. The formula is often written as follows:

$$\text{Position} = \frac{n+1}{2}.$$

The mistake may be due to the point that unlike the mean, a formula is not necessary to get the value of the median. It should be clear in the above examples that the value of the median is <u>not</u> the same as the position of the median.

Sample Problem and Solution: Given the following data set, rank the data, find the position of the median, and find the value of the median.

$$4.3, 3.8, 3.6, 4.2, 4.8, 2.1$$

Solution: Step 1. Rank the data: 2.1 3.6 3.8 4.2 4.3 4.8

Step 2. $\text{Position} = \frac{6+1}{2} = 3.5$

Step 3. The value halfway between 3.8 and 4.2 is 4.0.

Step 4. Median = 4.0

Sample Problem and Answer: Given the following data set, rank the data set, find the location and the value of the median.

$$8.1, 6.0, 5.7, 3.2, 7.1$$

Answer: Step 1. Ranked data: 3.2 5.7 6.0 7.1 8.1

Step 2. Position $= \dfrac{6}{2} = 3$.

Step 3. Median $= 6.0$

Sample Problem and Answer: Given the following data set, rank the data set, find the location and the value of the median.

$$6, 3, 5, 4, 8, 5$$

Answer: Step 1. Ranked data: 3 4 5 5 6 8

Step 2. Position $= 3.5$.

Step 3. Median $= 5.0$

We need to note that 5 is the value halfway between the third and fourth data values; which happen to be the same number in this example.

Hot Spot #2 – Effect Of Skewing On The Mean And Median

When we have a symmetric bell shaped distribution, the mean, median and mode are all located at the center of the distribution. However, the distribution is often skewed, meaning that one tail of the curve is missing. We say that the distribution is skewed to the side that has the tail of the curve. The following figures illustrate two skewed distributions.

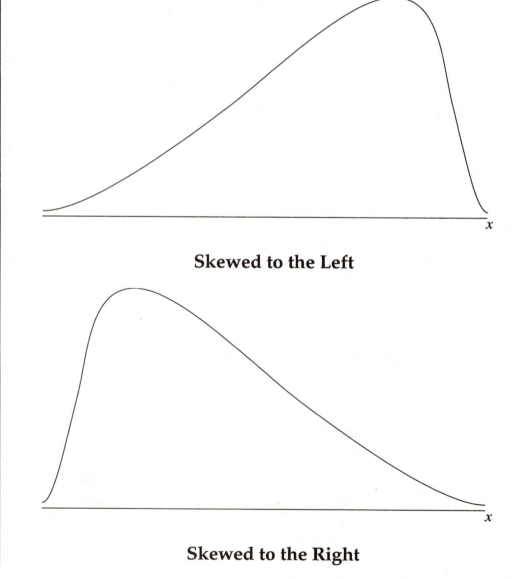

Figure 3-1:
Long Tail To The
Left (skewed)

Skewed to the Left

Figure 3-2:
Long Tail To The
Right (skewed)

Skewed to the Right

When a distribution is skewed to the left, there are more values at the lower end of the distribution. As a result both the mean and the median reflect the lower values. The following data set illustrates the idea. We will start with a symmetric bell shaped distribution and calculate the mean, median, and mode.

$$2, 3, 4, 4, 4, 5, 5, 5, 5, 5, 5, 6, 6, 6, 7, 8$$

The mode, median, and mean are all equal to 5 with

$$\bar{x} = \frac{\sum x}{n} = \frac{80}{16} = 5.$$

Now we will look at what happens with the skewed distribution we get when we throw out the top six data values. The distribution will now be skewed to the left.

$$2, 3, 4, 4, 4, 5, 5, 5, 5, 5$$

The mean has been shifted to the left; the same side as the skewing.

$$\bar{x} = \frac{\sum x}{n} = \frac{42}{10} = 4.2$$

We can also see what happens to both the mode and the median. The mode was not affected by the skewing; it is 5 for both data sets. But we find that the median, like the mean, reflects the skewing to the left with the median for the skewed data set $\tilde{x} = 4.5$. Both the mean and the median are < 5 and reflect the skewing. However, the mean was moved more to the left than the median, with $\bar{x} < \tilde{x}$, because the distribution was skewed to the left. In general the mean is more sensitive to extremes and will be moved further than the median to the skewed side of the distribution.

The following figures show the general relative positions of the mean, median and mode for skewed distributions:

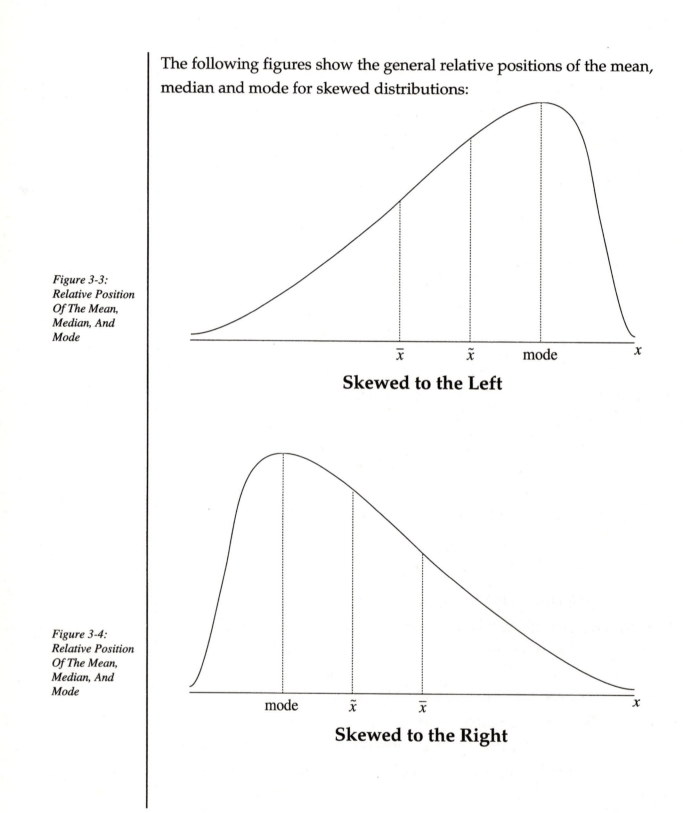

Figure 3-3:
Relative Position
Of The Mean,
Median, And
Mode

$\bar{x}$ $\tilde{x}$ mode x

Skewed to the Left

Figure 3-4:
Relative Position
Of The Mean,
Median, And
Mode

mode $\tilde{x}$ $\bar{x}$ x

Skewed to the Right

Hot Spot #3 – $\sum x^2$ Versus $\left(\sum x\right)^2$

The Greek letter sigma, Σ, is one of the most useful and important notations that we find in statistics. The symbol $\sum$ (sigma) is used to indicate a sum. The expression $\sum x$ means that we are to add up all the data values. Actually, the expression $\sum x$ is a shortened form of the following expression:

$$\sum_{i=1}^{n} x_i = x_1 + x_2 + x_3 + \cdots + x_n$$

The shortened form $\sum x$ is much easier to write and remember. The important point to remember is that whatever follows the sigma sign is to be added together. It may help to think or say "summation of x" rather than "sigma x" when we see $\sum x$.

For example, consider the following data set:

$$2, 8, 4, 9, 5, 6$$

If we want the sum of the data values, we write the expression $\sum x$ and find the summation of x as follows:

$$\sum x = 2 + 8 + 4 + 9 + 5 + 6 = 34$$

If we want to square the sum of the data values, we use a set of parentheses to indicate that the sum is found first, and we write the expression $\left(\sum x\right)^2$. We then square the sum as shown below:

$$\left(\sum x\right)^2 = (34)^2 = 1,156$$

If we want to find the sum of squares, we write the expression $\sum x^2$. We first square each data value. We then add the squares to get our answer:

$$\sum x^2 = 2^2 + 8^2 + 4^2 + 9^2 + 5^2 + 6^2$$
$$= 4 + 64 + 16 + 81 + 25 + 36$$
$$= 226$$

It should be very clear that the two summation expressions

$$\left(\sum x\right)^2 \text{ and } \sum x^2$$

do not give the same value. We use both of these expressions in the alternative formula for the variance. We will also use the expressions in the chapter near the end of the book on correlation and regression. It is important to know how to use the sigma notation correctly.

Hot Spot #3 Sample Problem and Solution: Find $\sum x$ and $\sum x^2$ for the following data set: 1, 2, 3, 4, 5

Solution: $\sum x = 1 + 2 + 3 + 4 + 5 = 15$

$\sum x^2 = 1^2 + 2^2 + 3^2 + 4^2 + 5^2 = 1 + 4 + 9 + 16 + 25 = 55$

Sample Problem and Answer: Find $\sum x$ and $\left(\sum x\right)^2$ for the following data set: $\qquad$ 1.1, 2.3, 1.2, 2.5, 1.0

Answer: $\sum x = 8.1$

$$\left(\sum x\right)^2 = (8.1)^2 = 65.61$$

Sample Problem and Answer: Find $\sum x$, $\sum x^2$, and $\left(\sum x\right)^2$ for the following data set: $\qquad$ -2, -1, 0, 1, 2

Answer: $\sum x = 0$

$$\sum x^2 = 10$$

$$\left(\sum x\right)^2 = (0)^2 = 0$$

Hot Spot #4 – $\sum f$ In The Formulas For $\bar{x}$ And s For Grouped Data

One of the most common mistakes that is made in calculating the mean and variance for grouped data is to divide by the number of intervals rather than the sample size. The formula for the mean for grouped data can be written in two ways:

$$\bar{x} = \frac{\sum mf}{\sum f} \qquad \text{or} \qquad \bar{x} = \frac{\sum mf}{n}$$

Surprisingly, it is the second form with the divisor of n that often creates the problem. It is easy to mistake the value of n as the number of intervals rather than the sample size. The first form, with the divisor $\sum f$, focuses on the sum of the frequencies.

The same problem exists with the formula for sample variance for grouped data. The formula given below is the alternative formula for sample variance:

$$s^2 = \frac{\sum m^2 f - \dfrac{\left(\sum mf\right)^2}{n}}{n-1} \qquad \text{where } n = \sum f$$

It may help to rewrite the formula as follows using $\sum f$ to again focus on the sum of the frequencies rather than the number of intervals when working with a grouped frequency distribution:

$$s^2 = \frac{\sum m^2 f - \dfrac{\left(\sum mf\right)^2}{\sum f}}{\sum f - 1}$$

Hot Spot #5 – Checking To See That We Have s Rather Than s^2

✳
*Mann
Section 3.2.2*

One of the easiest mistakes to make when calculating the standard deviation is to forget to take the square root. There is an easy way to check on an answer for the standard deviation of distributions that are approximately bell shaped. Generally speaking, the standard deviation should be a value between $\frac{R}{6}$ and $\frac{R}{4}$ where R is the range of the data.

- If the sample size is large, with $n \geq 30$, the standard deviation should be closer to $\frac{R}{6}$. The check value $\frac{R}{6}$ comes from the Empirical Rule where the distribution is divided into 6 equal regions that are each equal in width to the standard deviation.

- If the sample size is small, with $15 < n < 30$, the standard deviation should be closer to $\frac{R}{4}$. The $\frac{R}{4}$ is an adjustment for small sample sizes between 16 and 30 where we do not often get data from the tails of the curves.

- If the sample size is 15 or less, a check provided by Lincoln E. Moses of Stanford University should be used. The check value for samples smaller than 16 is defined as

$$\frac{R}{\sqrt{n}}$$

and is called the "poor man's standard deviation" (*p.m.s.d.*). Dr. Moses, in his book <u>Think And Explain With Statistics</u>, (Addison-Wesley Publishing Company, 1986), gives credit for the relationship between s and *p.m.s.d.* to Nathan Mantel.

Hot Spot #5 Sample Problem and Solution: Given the following set of data, calculate the sample standard deviation and check the value using one of the check values:

$$2, 7, 5, 13, 10, 4, 15, 2, 10, 2$$

Solution: Step 1. $\bar{x} = \dfrac{\sum x}{n} = \dfrac{70}{10} = 7$

Step 2. $s^2 = \dfrac{\sum (x - \bar{x})^2}{n-1} = \dfrac{206}{9} = 22.89$

Step 3. $s = 4.78$

Step 4. $n = 10 < 16$; use $s \approx \dfrac{R}{\sqrt{n}} = \dfrac{15-2}{\sqrt{10}} = \dfrac{13}{3.16} = 4.11$

Step 5. The check value agrees with the calculated value of s; note the large difference between the value of s^2 and the check value.

Sample Problem and Answer: The following values were calculated from a data set containing 35 values and having a range of 20. Check the value of the sample standard deviation using one of the check values:

$$n = 35 \qquad\qquad s = 12.2 \qquad\qquad R = 20$$

Answer: Step 1. $n = 35 > 15$; use the $\frac{R}{6}$ and $\frac{R}{4}$ values.

Step 2. $\dfrac{R}{6} = \dfrac{20}{6} = 3.3\overline{3}$ and $\dfrac{R}{4} = \dfrac{20}{4} = 5.0$

Step 3. $s = 12.2$ is not in the interval $(3.3\overline{3}, 5.0)$. This suggests that the correct value of $s = \sqrt{12.2} = 3.49$.

Sample Problem and Answer: Find the sample standard deviation for the following data set and check the value using one of the check values:

$$2, 3, 7, 3, 5$$

Answer: Step 1. $\bar{x} = \dfrac{20}{5} = 4$

Step 2. $s^2 = \dfrac{16}{4} = 4$

Step 3. $s = 2$

Step 4. Use $\dfrac{R}{\sqrt{n}} = \dfrac{5}{2.236} = 2.236$

Step 5. The check value is comparable to the calculated value of 2: Note the large difference between 2.236 and 4.

Hot Spot #6 – The Meaning Of $\bar{x} \pm ks$

Both the Empirical Rule and Chebyshev's theorem require that we create a symmetric interval about the mean. We use the phrase, $\bar{x} \pm ks$, which reads as the mean plus or minus k standard deviations:

where k is a 1, 2, or 3 for the Empirical Rule
and $k > 1$ for Chebyshev's theorem.

For example, if $k = 2$, then we have the interval $(\bar{x} - 2s, \bar{x} + 2s)$. If $\bar{x} = 70$ and $s = 10$ for a distribution of test scores, then the interval for the mean plus or minus two standard deviations would be from 50 to 90.

$$\bar{x} - 2s = 70 - (2)(10) = 70 - 20 = 50$$

$$\text{and}$$

$$\bar{x} + 2s = 70 + (2)(10) = 70 + 20 = 90$$

Mann
Sections
3.4.1 And 3.4.2

If we wanted the mean plus or minus three standard deviations, we would write the expression that follows:

$$\bar{x} \pm 3s$$

$$\bar{x} - 3s \text{ and } \bar{x} + 3s$$

$$\bar{x} - 3s = 70 - (3)(10) = 70 - 30 = 40$$
$$\text{and}$$
$$\bar{x} + 3s = 70 + (3)(10) = 70 + 30 = 100$$

When we create the intervals for Chebyshev's theorem we often use non-integer values of k. For example, we may want to find the mean plus or minus 1.5 standard deviations. The expression would be written as:

$$\bar{x} \pm 1.5s$$

Using our test example with $\bar{x} = 70$ and $s = 10$, the interval involves the following steps:

$$\bar{x} - 1.5s = 70 - (1.5)(10) = 70 - 15 = 55$$
$$\text{and}$$
$$\bar{x} + 1.5s = 70 + (1.5)(10) = 70 + 15 = 85$$

Hot Spot #6 Sample Problem and Solution: Find the interval for the sample mean plus or minus one standard deviation for the above example of test scores. The mean is equal to 70 and the standard deviation is equal to 10.

Solution: $\bar{x} \pm 1s = 70 \pm 10$

$$\bar{x} - 1s = 70 - (1)(10) = 60$$

$$\bar{x} + 1s = 70 + (1)(10) = 80$$

Sample Problem and Answer: Find the interval for $\bar{x} \pm 2.5s$ for the distribution with a sample mean of 1,200 and sample standard deviation of 50.

Answer: The interval goes from 1,075 to 1,325.

Sample Problem and Answer: Find the interval for $\bar{x} \pm 3s$ for a distribution with sample mean of 0.5 and sample standard deviation of 0.125.

Answer: The interval goes from 0.125 to 0.875.

Hot Spot #7 – Understanding Chebyshev's Theorem

Unlike the Empirical Rule that can only be used with symmetric bell shaped distributions, Chebyshev's theorem can be used with any distribution. There are two distinct problem areas with Chebyshev's theorem. The first is the difficulty with the phrase "at least" and the second is with the formula $1 - \frac{1}{k^2}$. The theorem states that **at least** $1 - \frac{1}{k^2}$ percent of the data will be within k standard deviations of the mean. It may help if we break the process into four steps as follows:

Step 1. Determine the interval $\bar{x} \pm ks$

Step 2. Find the value of $1 - \dfrac{1}{k^2}$

Step 3. Change the value in step 2 to a percent.

Step 4. Write the statement that at least the percent of data found in step 3 is in the interval found in step 1.

One of the major concerns that we will have as we continue with our study of statistics is what happens when our assumptions about a bell shaped distribution are violated. Chebyshev's theorem provides us with an alternative analysis that we can use. The results of the analysis are not as strong as what could be said if the distribution were bell shaped. But the theorem does allow us to make a statement about the results even though our original assumption was rejected. The last chapter of the text deals exclusively with forms of analysis that are possible when assumptions about the distribution are violated. The generic term for these procedures is non-parametric statistics. Chebyshev's theorem is the first of many such adjustments we may be required to make if the assumption of a symmetric bell shaped distribution is violated. In this case Chebyshev's theorem gives us a worst-case scenario.

As an example, we will consider a distribution of test scores that are badly skewed to the right with a sample mean of 80 and a sample standard deviation of 5. If $k = 2$, the theorem tells us that at least 75% of the scores will fall in the interval from 70 to 90. The following steps show how the values were found:

Step 1. Determine the interval $\bar{x} \pm ks$

$$\bar{x} \pm 2s = 80 \pm (2)(5) = 80 \pm 10$$

The interval is 70 to 90.

Step 2. Find the value of $1 - \dfrac{1}{k^2}$

$$1 - \frac{1}{2^2} = 1 - \frac{1}{4} = \frac{3}{4}$$

Step 3. Change the value in step 2 to a percent.

$$\frac{3}{4} = 75\%$$

Step 4. Write the statement that at least the percent of data found in step 3 is in the interval found in step 1.

At least 75% of the data is found in the interval from 70 to 90.

It is important to understand that Chebyshev's theorem is also valid with symmetric bell shaped distributions. The percent of data in the interval is just a minimum statement. For example, if the data in the above example came from a symmetric bell shaped distribution, the Empirical Rule would tell us that 95% of the data would be in the interval from 70 to 90. In contrast, Chebyshev's theorem told us that at least 75% of the data was in the interval from 70 to 90. The important point to see is that 95% is consistent with the statement that at least 75% of the data was in the interval from 70 to 90. The added information about the distribution being bell shaped allows us to tighten up our prediction about how much data is in the interval. If we know the distribution is bell shaped we can give a specific value of 95%. But if the distribution is not bell shaped, all we can do is give a minimum value of 75%.

A very common error is made when using Chebyshev's theorem. Rather than stating that at least 75% of the data is in the interval of the mean plus or minus 2 standard deviations, a statement is made that 75% of the data is in the interval. This is a serious mistake and should be avoided. Remember to use the phrase "at least" when using Chebyshev's theorem.

One final word about Chebyshev's theorem is in order before looking at the sample problems. Although understanding the theorem may take some effort, the level of thinking and analysis that is used in working with Chebyshev's theorem is both very necessary and essential to understanding inferential statistics.

Hot Spot #7 Sample Problem and Solution: Use Chebyshev's theorem to find the percent of the data within 3 standard deviations of the sample mean for $\bar{x} = 55$ and $s = 10$.

Solution: Step 1. $\bar{x} \pm 3s = 55 \pm 30$ gives the interval 25 to 85

Step 2. $1 - \dfrac{1}{k^2} = 1 - \dfrac{1}{3^2} = 1 - \dfrac{1}{9} = \dfrac{8}{9}$

Step 3. $\dfrac{8}{9} \approx 89\%$

Step 4. At least 89% of the data is in the interval from 25 to 85.

Sample Problem and Answer: Use Chebyshev's theorem to find the minimum percent of data within 1.5 standard deviations of the sample mean.

Answer: $1 - \dfrac{1}{1.5^2} = 1 - \dfrac{1}{\left(\frac{3}{2}\right)^2} = 1 - \dfrac{4}{9} = \dfrac{5}{9} \approx 56\%$

At least 56% of the data is in the interval $\bar{x} \pm 1.5s$.

Sample Problem and Answer: Assume that a distribution of data is non-symmetric with a sample mean of 70 and sample standard deviation of 10. Find the minimum percent of the data within 4 standard deviations of the mean and give the interval.

Answer: The interval goes from 30 to 110.

$$1 - \dfrac{1}{k^2} = 1 - \dfrac{1}{4^2} = 1 - \dfrac{1}{16} = \dfrac{15}{16} \approx 94\%$$

At least 94% of the data is in the interval from 30 to 110.

Hot Spot #8 – When P_k Does Not Result In k% Of The Data Below P_k

When we work with the median of a distribution, it is common to argue that half of the data values are below the median and half of the data values are above the median. But as is often the case, we find sets of data that make the above interpretation false. The following data set illustrates the problem:

$$2, 3, 4, 5, 5, 6, 7, 7, 8$$

There are nine data values in the set, and the median is 5. Only three of the data values are lower than the median; whereas, four of the values are above the median. Clearly, in this example, the median of 5 does not separate the data into two sets of equal size. It is because of this type of problem that we define the median as the value that is in the middle of the ranked data list.

A very similar problem occurs when we work with percentiles. The kth percentile, P_k, is supposed to be a value that separates the

bottom k% of the data from the top $(100 - k)$% of the data. For example, the median is sometimes defined as P_{50}, the 50*th* percentile. That is the median should separate the set of data into two subsets, each containing 50% of the data.

As we see from the above example, the median of 5 does not separate the data as we would like it to. This is often a problem with small sets of data, but it becomes less of a problem as the data sets get larger. Sometimes we try to get around the problem by saying that the kth percentile is the value at which k% of the data is at that value or below it. However in the above data set five of the data values are less than or equal to P_{50}. This gives us 56% rather than the desired 50%. Simply stated, there are times when it is impossible to define percentile in a way that avoids inconsistencies.

As with the median, we must be very careful not to confuse the location of the percentile with the value of the percentile. For example, in the above data set the 25*th* percentile is found in the 3*rd* position. We find the position of P_{25} using the formula:

$$\text{Position of } P_k = \left(\frac{kn}{100}\right)$$

To avoid the confusion of position and value, we can get around the problem by writing out the position of P_{25} as follows:

Position of the twenty-fifth percentile $= (0.25)(9) = 2.25$.

We approximate the value of the 2.25 term by averaging the 2nd and 3rd terms in the arranged data as follows:

$$\frac{3+4}{2} = 3.5$$

We reserve the symbol P_{25} for the value of the twenty-fifth percentile as follows:

$$P_{25} = 3.5$$

Hot Spot #9 – Rounding Off

One is often tempted to round off the results of calculations like the mean and standard deviation to one or two digits beyond the number of digits after the decimal point in the original data. For example, if the original data values were whole numbers, one usually rounds off the answers to the nearest tenth. This is particularly the case when a calculator is used with a formula to do the calculations for the mean and standard deviation. The practice of rounding off can create difficulties in two ways.

The first problem occurs if one uses the following formula to find the standard deviation:

$$s = \sqrt{\frac{\sum(x - \bar{x})^2}{n - 1}}$$

If the mean is a decimal value that goes beyond two digits after the decimal point, rounding off to the nearest tenth will cause the value of the standard deviation to be incorrect. The error may appear to be small, but the effect on a subsequent value like the standard score is quite significant.

In later chapters we use what is called the standard score (z value) to do hypothesis testing where an assumed value of a population parameter is challenged. Large z values result in a decision to reject the assumption about the population parameter. For example, in hypothesis tests about μ, the standard score is defined as follows:

$$z = \frac{\bar{x} - \mu}{\frac{\sigma}{\sqrt{n}}}$$

where the value of s (sample standard deviation) is substituted for σ.

If s is in error, the resulting z value will be in error. The decision we then make could also be in error simply because of rounding off.

This then brings up the question of when is it all right to round off. The answer to the question involves two rules:

Rounding Off Rules

Rule I: If we are simply reporting a value such as the mean or standard deviation for a report, then rounding off at the end of each calculation is permissible. As a rule of thumb, round off to one more digit after the decimal place than in the original data.

Rule II: If the value calculated is to be used in a subsequent calculation, the value should not be rounded off. As a rule of thumb, one should use as many digits as possible in any calculation that uses the value just obtained. Fortunately calculators allow us to carry as many as 10 digits in any calculations we do. The mean should not be rounded off when it is used to find the standard deviation. The standard deviation should not be rounded off when it is used to find the standard score.

CHAPTER 3 DISCUSSION QUESTIONS

These questions may be used in your study group or simply as topics for individual reflection. Whichever you do, take time to explain verbally each topic to insure your own understanding. Since these questions are intended as topics for discussion, answers to these questions are not provided. If you find that you are not comfortable with either your answers or that your group has difficulty with the topic, take time to meet with your professor to get help.

1. When is it better to use the median rather than the mean?

2. Discuss the difference between the Empirical Rule and Chebyshev's Theorem.

3. Why is the variance a better measure of dispersion than the range?

4. What is meant by the phrase, the mean plus or minus k standard deviations?

5. Why is the mean for a grouped frequency distribution usually different than the mean for the original data?

6. When is it better to use the alternative formulas for calculating the variance and standard deviation?

7. What is the difference between $\sum x^2$ and $\left(\sum x\right)^2$?

8. How are the mean and median affected by skewing?

9. What is the difference between a parameter and a statistic?

10. What is the difference between the value of the median and the position of the median?

CHAPTER 3 TEST

Use the following sample data for problems 1 through 5:

$$4, 2, 3, 5, 2, 8, 9, 4, 5, 8$$

1. Find the position of the median and the value of median.

2. Find the mean.

3. Find the range.

4. Find the standard deviation.

5. Find the first quartile.

Use the following information for problems 6 through 8:
$\bar{x} = 70$, $s = 5$; the sample is from a normal population.

6. What percent of the data is in the interval from 60 to 80?

7. What are the values for the interval $\bar{x} = \pm 3s$?

8. Using Chebyshev Theorem what is the percent of the data in the interval $\bar{x} = \pm 3s$.

Use the following grouped frequency table for problems 9 and 10.

Class Boundaries	Frequency	Class Marks
44.5–55.5	2	50
55.5–66.5	5	60
66.5–77.5	3	70

9. Find the mean.

10. Find the variance.

CHAPTER 3 TEST Questions and Answers

Use the following sample data for problems 1 through 5:

$$4, 2, 3, 5, 2, 8, 9, 4, 5, 8$$

1.   Find the position of the median and the value of median.

Answer: The data, in ranked order, are 2, 2, 3, 4, 4, 5, 5, 8, 8, 9.

$$\text{Position of median} = \frac{n+1}{2} = \frac{11}{2} = 5.5.$$

The median is the value halfway between 4 and 5.

$$\text{Median} = \frac{4+5}{2} = 4.5.$$

2.   Find the mean.

Answer: $\bar{x} = \dfrac{\sum x}{n} = \dfrac{50}{10} = 5$

3.   Find the range.

Answer: $R = H - L = 9 - 2 = 7$

4.   Find the standard deviation.

Answer:

$$s = \sqrt{\frac{n\sum x^2 - \left(\sum x\right)^2}{n(n-1)}} = \sqrt{\frac{10(308) - (50)^2}{9(10)}} = \sqrt{\frac{580}{90}} = \sqrt{6.\overline{4}} = 2.538591$$

5.   Find the first quartile.

Answer: The first quartile is $Q_1 = P_{25}$.

The position of P_{25} is $\left(\dfrac{25}{100}\right)10 = 2.5$. Since 2.5 is not a whole

number, we use the average of the *2nd* and *3rd* terms of the

data: $\dfrac{2+3}{2} = 2.5$

Coincidentally, the average equals the position this time.

$$Q_1 = P_{25} = 2.5$$

Use the following information for problems 6 through 8:

$\bar{x} = 70,\ s = 5$; the sample is from a normal population.

6.   What percent of the data is in the interval from 60 to 80?

Answer: The interval is $\bar{x} \pm 2s$. Using the Empirical Rule, 95% of the
data is in the interval from 60 to 80.

7.   What are the values for the interval $\bar{x} = \pm 3s$?

Answer: $\bar{x} = \pm 3s = 70 \pm 3(5) = 70 \pm 15$

The interval is from 55 to 85.

8.   Using Chebyshev Theorem what is the percent of the data in the
interval $\bar{x} = \pm 3s$.

Answer: At least $1 - \dfrac{1}{3^2} = \dfrac{8}{9} = 88.9\%$

Use the following grouped frequency table for problems 9 and 10.

Class Boundaries	Frequency	Class Marks
44.5–55.5	2	50
55.5–66.5	5	60
66.5–77.5	3	70

9. Find the mean.

Answer: $\bar{x} = \dfrac{\sum mf}{n} = \dfrac{2(50) + 5(60) + 3(70)}{10} = \dfrac{610}{10} = 61$

10. Find the variance.

Answer: $s^2 = \dfrac{n\left(\sum m^2 f\right) - \left(\sum mf\right)^2}{n(n-1)}$

$= \dfrac{10\left(2500(2) + 3600(5) + 4900(3)\right) - (610)^2}{10(9)}$

$= \dfrac{10(37,700) - 372,100}{90}$

$= \dfrac{4900}{90}$

$= 54.4\overline{4}$

$\sum\circ$

chapter four

PROBABILITY

*This branch of mathematics [probability] is the
only one, I believe,
in which good writers frequently get results
entirely erroneous.*
— Charles Sanders Pierce

INTRODUCTION

This chapter on probability is often the most challenging chapter
in the introductory statistics course. In many respects, one could do a
minimal amount of work in probability and still be able to do most of
the problems in statistics. This, of course, brings us to the question of
"Why do we have this chapter on probability?"

In part, the answer to the above question involves the creation
and development of the introductory statistics course in the 1960's. At
that time, the usual statistics curriculum was taught in upper division
and involved a semester of probability; followed by a second semester of
statistics. When the lower division statistics course evolved, a chapter
on probability was considered to be an important part of the course.

Aside from the historical perspective, there is good reason to
understand the ideas in the probability chapter. Our early work in
constructing relative frequency histograms and polygons from the
relative frequencies of grouped data provided us with what are called
experimental or *a posteriori* probabilities. It only seems natural to extend

this idea in a formal way to theoretical or *a priori* probabilities.

Theoretical probabilities are developed using the idea of a **sample space** that results from the outcomes of the experiment. We will focus on the outcomes, called **simple events**. Capital letters, such as *A* and *B* are used to represent **events** that are collections or sets of the simple outcomes.

Events are classified in a variety of ways using terms that have an intuitive appeal that helps us understand their meaning. For example, we will talk about events being **independent, mutually exclusive**, and **complementary**. We then find the probability of events occurring using the idea of counting. Unfortunately, the counting process can become cumbersome. It is simplified through the use of **Tree Diagrams, Venn Diagrams**, the **Counting Rule** and formulas that focus on the probability of two events occurring. Specifically, we will be interested in the probability of the **compound events** (*A* and *B*) and (*A* or *B*), and the **conditional probability** of *A* given *B*.

We then apply the concepts and formulas by doing problems that involve experiments such as tossing coins, drawing cards, and throwing dice. The problems provide us an opportunity to become familiar with the terms, formulas, and notation used in probability.

Many of the ideas we develop in this chapter are essential to understanding the work we do in the chapters that follow. For example, in the next two chapters, we use the concept of probability in defining both discrete and continuous probability distributions. These theoretical distributions are used as models for the data distributions we worked with earlier.

We use the distributions in the hypothesis testing that is such an important part of inferential statistics. In fact, the very essence of hypothesis testing involves finding the probability of getting the value of the sample statistic we use to challenge the accepted value of the population parameter.

CHAPTER 4 HOT SPOTS

1. **Using Tree Diagrams.**
   Starts on **page 4–5**. Problems on **pages 4–10, 4–11, 4–12.**
   Mann 4.1

2. **Using The Counting Rule.**
   Starts on **page 4–12**. Problems on **pages 4–13, 4–14.** Mann 4.3

3. **Using The Complement.**
   Starts on **page 4–14**. Problems on **page 4–16.** Mann 4.7

4. **Independent Events Versus Dependent Events.**
   Starts on **page 4–17**. Problems on **pages 4–19, 4–20.** Mann 4.6, 4.8.2

5. **Mutually Exclusive Versus Independent Events.**
   Starts on **page 4–20.** Mann 4.5, 4.6, 4.8.2

6. **Special Vs. General Formula For** $P(A \text{ and } B)$ **&** $P(A \text{ or } B)$.
   Starts on **page 4–21.** Mann 4.8.2, 4.9.2

7. **Deciding On Which Formula To Use.**
   Starts on **page 4–22**. Problems on **pages 4–23, 4–24.**
   Mann 4.4, 4.5, 4.6, 4.7, 4.8, 4.9

8. $P(A \text{ and } B)$ **Versus** $P(A|B)$.
   Starts on **page 4–25.** Mann 4.8.2

If you find other HOT SPOTS, write them down and use them as a focus of your discussions in the study group. Or you can use the HOT SPOT as the topic for a help session with your professor.

USE THIS PAGE FOR KEEPING TRACK OF TERMS AND NOTATION

Hot Spot #1 – Using Tree Diagrams

We will start with the Tree Diagram. It is perhaps the most important topic in this chapter. The Tree Diagram is very similar to a factor tree that we first see in the prime factoring of composite numbers. For example, in factoring 35 we get the following factoring tree:

✳
Mann
Section 4.1

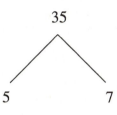

35

5   7

Figure 4–1:
5 × 7 Factor
Tree

The factor tree branches out to the two possible factors. We consider the toss of one coin to be like starting with a number. The two outcomes of H (heads) and T (tails) are like the factors. We diagram the outcomes just as we did the factoring:

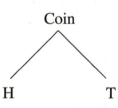

Coin

H   T

Figure 4–2:
Coin Toss
Diagram

We usually turn the Tree Diagram on its side as illustrated below:

Tree Diagram

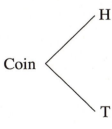

Coin

H

T

Figure 4–3:
Coin Toss Tree
Diagram

We can treat the toss of two coins like the factoring of 36. The second factoring resembles the second toss of the coin. The list of factors in the last step is like the list of outcomes we get in the tree diagram. The two diagrams are illustrated below:

Factor Tree

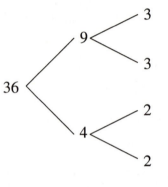

Figure 4–4:
Factor Tree

Tree Diagram

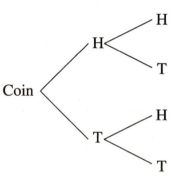

Figure 4–5:
Tree Diagram

We now complete the Tree Diagram by listing the probability of each outcome on the appropriate branch:

Tree Diagram

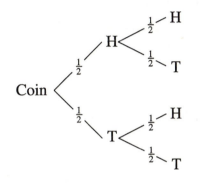

Figure 4–6:
Tree Diagram
w/Probabilities

The Tree Diagram can be used to help us list both the outcomes of an experiment and the probability of each of the outcomes:

Tree Diagram

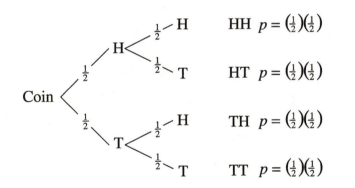

Figure 4–7:
Tree Diagram
w/Outcomes

There are times when all we do is list outcomes without writing in the probabilities of each outcome. In fact, we may not know the probabilities. We then simply use the Tree Diagram as a tool to help us understand how to do a problem. The following Tree Diagram, without the probabilities written in, illustrates the toss of one die (the singular of "dice"):

Tree Diagram

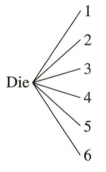

Figure 4–8:
Tree Diagram,
Toss Of One Die

If we were to toss the die a second time the Tree Diagram would be more difficult to draw and require more space. It is drawn on the following page as a solution to the first **Sample Problem and Solution**. The solution will take a full page to draw. This time, neither the probabilities nor the outcomes of the experiment will be listed.

The Tree Diagram is perhaps most useful when we are illustrating probability experiments that involve drawing without replacement. When we include the probabilities for each branch and the list of outcomes for the experiment, the diagram provides us the answers for questions about conditional probability.

As an example, we will draw two cards without replacement from a regulation deck containing 52 cards with four suits and four aces. We are interested in drawing an ace. Each time we draw, we will either get an ace (A) or not an ace ($\overline{A}$). The probability of drawing A on the first draw is $\frac{4}{52}$. But the probability of A on the second draw depends on what happened on the first draw. The probability of $\overline{A}$ on the first draw is $\frac{48}{52}$. But again, the probability of $\overline{A}$ on the second draw depends on what happened on the first draw. The following Tree Diagram helps us organize what is starting to sound complicated:

Tree Diagram

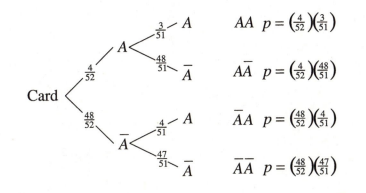

$$AA \quad p = \left(\tfrac{4}{52}\right)\left(\tfrac{3}{51}\right)$$

$$A\overline{A} \quad p = \left(\tfrac{4}{52}\right)\left(\tfrac{48}{51}\right)$$

$$\overline{A}A \quad p = \left(\tfrac{48}{52}\right)\left(\tfrac{4}{51}\right)$$

$$\overline{A}\,\overline{A} \quad p = \left(\tfrac{48}{52}\right)\left(\tfrac{47}{51}\right)$$

Figure 4–9: Tree Diagram, Draw of Two Cards

Hot Spot #1 Sample Problem and Solution: Draw the Tree Diagram for two tosses of a six sided die.

Solution:

Tree Diagram

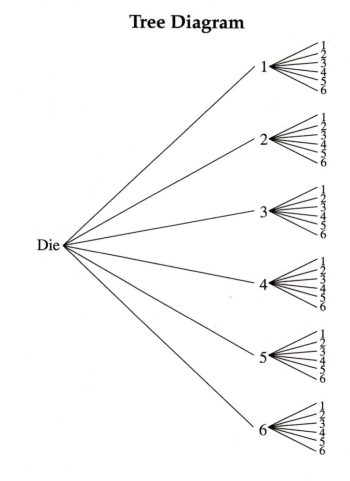

Figure 4–10:
Tree Diagram,
Two Tosses Of A
Die

A Study Guide For Statistics

Sample Problem and Answer: Draw a Tree Diagram for three tosses of a coin. Include the probabilities and the outcomes of the experiment.

Answer:

Tree Diagram

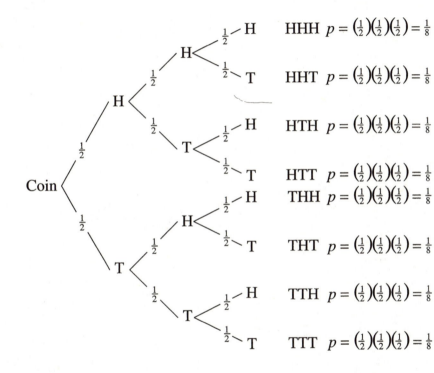

$$\text{HHH} \quad p = \left(\tfrac{1}{2}\right)\left(\tfrac{1}{2}\right)\left(\tfrac{1}{2}\right) = \tfrac{1}{8}$$

$$\text{HHT} \quad p = \left(\tfrac{1}{2}\right)\left(\tfrac{1}{2}\right)\left(\tfrac{1}{2}\right) = \tfrac{1}{8}$$

$$\text{HTH} \quad p = \left(\tfrac{1}{2}\right)\left(\tfrac{1}{2}\right)\left(\tfrac{1}{2}\right) = \tfrac{1}{8}$$

$$\text{HTT} \quad p = \left(\tfrac{1}{2}\right)\left(\tfrac{1}{2}\right)\left(\tfrac{1}{2}\right) = \tfrac{1}{8}$$

$$\text{THH} \quad p = \left(\tfrac{1}{2}\right)\left(\tfrac{1}{2}\right)\left(\tfrac{1}{2}\right) = \tfrac{1}{8}$$

$$\text{THT} \quad p = \left(\tfrac{1}{2}\right)\left(\tfrac{1}{2}\right)\left(\tfrac{1}{2}\right) = \tfrac{1}{8}$$

$$\text{TTH} \quad p = \left(\tfrac{1}{2}\right)\left(\tfrac{1}{2}\right)\left(\tfrac{1}{2}\right) = \tfrac{1}{8}$$

$$\text{TTT} \quad p = \left(\tfrac{1}{2}\right)\left(\tfrac{1}{2}\right)\left(\tfrac{1}{2}\right) = \tfrac{1}{8}$$

Figure 4–11:
Tree Diagram,
Three Tosses Of
A Coin

Sample Problem and Answer: Draw the Tree Diagram for the experiment consisting of tossing one coin followed by the draw of one card from a deck where success is drawing an ace. Include the individual probabilities for each branch, the outcomes of the experiment, and their probabilities in the Tree Diagram.

Answer:

Tree Diagram

Figure 4–12:
Tree Diagram,
Toss Of One
Coin And Draw
Of One Card

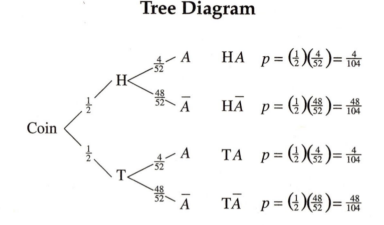

$$\text{HA} \quad p = \left(\tfrac{1}{2}\right)\left(\tfrac{4}{52}\right) = \tfrac{4}{104}$$

$$\text{H}\overline{\text{A}} \quad p = \left(\tfrac{1}{2}\right)\left(\tfrac{48}{52}\right) = \tfrac{48}{104}$$

$$\text{TA} \quad p = \left(\tfrac{1}{2}\right)\left(\tfrac{4}{52}\right) = \tfrac{4}{104}$$

$$\text{T}\overline{\text{A}} \quad p = \left(\tfrac{1}{2}\right)\left(\tfrac{48}{52}\right) = \tfrac{48}{104}$$

Mann
Section 4.3

Hot Spot #2 – Using The Counting Rule

On the surface, probability seems straight forward. After all, we only need to list the sample space and count which outcomes are a success. Unfortunately, listing outcomes is not always easy; even with a Tree Diagram. For example, the popular dice game, YAHTZEE, involves tossing five dice. This is the same as taking one die and tossing it five times. When we get all five dice with the same number we have a YAHTZEE! It is interesting to find the probability of a YAHTZEE, because it is unreasonable to list the outcomes using a Tree Diagram. Yet, it is easy to find the answer.

What we do is find a way of determining the number of outcomes without counting them. We know that there are only six ways of getting a YAHTZEE. All we need is the number of total possible outcomes to write the correct probability.

In the first **Sample Problem and Solution**, (**Hot Spot #1**) we found that there were 6×6, or 36, outcomes when we used two dice. All we have to do is realize that for every new dice we toss we have six new outcomes. The six new outcomes will be multiplied by the previous six outcomes. When we toss five dice we will have $6 \times 6 \times 6 \times 6 \times 6$, or 7,776 total possible outcomes. The probability of a YAHTZEE is $\frac{6}{7,776}$.

$$P(\text{YAHTZEE}) = \frac{6}{7,776}$$

The principle we just used is called the counting rule. Let us assume we do two things at once where there are k ways of doing the first thing and m ways of doing the second thing. There are then km ways of doing the two things at the same time. If we extend to n ways for doing a third thing, then we would have kmn ways of doing the three things together. The counting rule gives us a way of often finding the number of total outcomes without actually counting or listing the outcomes with a Tree Diagram.

The Counting Rule

Hot Spot #2 Sample Problem and Solution: How many outcomes are there when we toss 5 coins?

Solution: There are two outcomes for each coin. Since we are tossing five coins, we multiply $2 \times 2 \times 2 \times 2 \times 2$ to get 32 outcomes.

Sample Problem and Answer: We toss a six sided die and draw a card from a regulation deck of 52 cards where we record the suit drawn. How many outcomes are possible?

Answer: $6 \times 4 = 24$ outcomes.

Sample Problem and Answer: An experiment involves tossing a coin, tossing a six sided die, and drawing a card from a deck of 52 cards and recording the suit. How many possible outcomes?

Answer: $2 \times 6 \times 4 = 48$ outcomes.

✳
*Mann
Section 4.7*

Hot Spot #3 – Using The Complement

The YAHTZEE problem we did earlier in **Hot Spot #2** provides us with another insight regarding complementary events. The complement of an event A is everything in the sample space after A is removed. Since there were only six ways of getting a YAHTZEE, there are $7,776 - 6 = 7,770$ outcomes in the complement of the set of six outcomes that gave us a YAHTZEE. With this information, we can find the probability of not getting a YAHTZEE.

$$P(\text{Not a YAHTZEE}) = \frac{7,770}{7,776}$$

A simpler way of finding the probability of "not a YAHTZEE" is to use the probability formula for complementary events:

$$P(\overline{A}) = 1 - P(A)$$

Complementary Events Formula

Using the above formula,

$$P(\overline{Y}) = 1 - P(Y) = 1 - \frac{6}{7,776} = \frac{7,770}{7,776}$$

The probability formula for complementary events is very useful when the problem we are trying to do has many outcomes. The complement of the event then involves few outcomes. Finding the probability of at least one success is one of the classic problems using the idea of a complement. For example, suppose we are tossing a coin three times and we want the probability of at least one tail. We use the complement of at least one tail which is getting zero tails.

$$P(\text{at least one tail}) = 1 - P(\text{zero tails})$$

There is only one way of getting zero tails, namely getting exactly three heads. The answer is easy to set up as follows:

$$P(\text{at least one tail}) = 1 - \left(\tfrac{1}{2}\right)\left(\tfrac{1}{2}\right)\left(\tfrac{1}{2}\right) = 1 - \tfrac{1}{8} = \tfrac{7}{8} = 0.875$$

The key phrase that most often helps us know when to use the complement is "find the probability of at least one success."

Hot Spot #3 Sample Problem and Solution: In a toss of a die three times, what is the probability of getting at least one 3?

Solution: The complement of getting at least one 3 is to get zero 3's. This can be done in only one way. The probability of not getting a 3 is $\frac{5}{6} = 0.83\overline{3}$.

$$P(\text{at least one 3}) = 1 - P(\text{zero 3's})$$

$$= 1 - \left(\tfrac{5}{6}\right)\left(\tfrac{5}{6}\right)\left(\tfrac{5}{6}\right) = 1 - 0.58 = 0.42$$

Sample Problem and Answer: In a production process ten items are selected from a shipment. Assume that each item has a probability of 0.05 of being defective. What is the probability that at least one item will be defective?

Answer: We note that $P(\text{zero defectives}) = 0.95$ for each selection.

$$P(\text{at least one defective}) = 1 - P(\text{zero defectives}) = 1 - (0.95)^{10}$$

$$= 1 - 0.599 = 0.401$$

Sample Problem and Answer: In five tosses of a coin what is the probability of at least two heads?

Answer: The probability of at least two heads tells us we are interested in $x \geq 2$, where x is the number of heads. The complement of $x \geq 2$ is $x \leq 1$. Thus the complement of at least two heads is the event of getting either zero heads or one head. The

$P(\text{zero heads}) = \left(\frac{1}{2}\right)^5$ and the $P(\text{one head}) = 5\left(\frac{1}{2}\right)^5$. If we looked at a Tree Diagram for five tosses of a coin, we would find that five of the thirty-two branches have exactly one head.

$$
\begin{aligned}
P(\text{at least two heads}) \quad &= 1 - \left(P(\text{zero heads}) + P(\text{one head})\right) \\
&= 1 - \left(\left(\tfrac{1}{2}\right)^5 + 5\left(\tfrac{1}{2}\right)^5\right) = 1 - \left(\tfrac{1}{32} + \tfrac{5}{32}\right) \\
&= 1 - \tfrac{6}{32} = 1 - 0.1875 = 0.8125
\end{aligned}
$$

Hot Spot #4 – Independent Events Versus Dependent Events

The work with conditional probability leads us to the idea of independent events versus dependent events. The idea of independence suggests that one event does not influence a second event. However, the concept is really defined using a formula irrespective of whether or not the independence appeals to us intuitively. The definition for independent events A and B requires the following to be true:

$$P(A|B) = P(A) \text{ and } P(B|A) = P(B)$$

The two statements are equivalent. For example, If we know $P(A|B) = P(A)$, then we know $P(B|A) = P(B)$. If $P(A|B) \neq P(A)$ or $P(B|A) \neq P(B)$, the events are dependent.

We can illustrate the idea with the following example that involves tossing a six sided die. Find the probability that we get a 3, given we know the outcome is a multiple of 3.

$$\text{Let } A = \{3\} \qquad B = \{3,6\} \qquad S = \{1,2,3,4,5,6\}$$

✳
Mann
Sections
4.6 And 4.8.2

Independent
Events

$$P(A|B) = \frac{P(A \text{ and } B)}{P(B)} = \frac{\frac{1}{6}}{\frac{2}{6}} = \frac{1}{2} \qquad\qquad P(A) = \frac{1}{6}$$

$$P(A|B) \neq P(A)$$

So A and B **are not** independent events; they are dependent events.

In contrast, we will consider a similar problem again using a six sided die. This time we want the probability of an even number, given we know the outcome is a multiple of three.

$$\text{Let } A = \{2,4,6\} \quad B = \{3,6\} \qquad S = \{1,2,3,4,5,6\}$$

$$P(A|B) = \frac{P(A \text{ and } B)}{P(B)} = \frac{\frac{1}{6}}{\frac{2}{6}} = \frac{1}{2} \qquad\qquad P(A) = \frac{1}{2}$$

$$P(A|B) = P(A)$$

So A and B **are** independent events.

We need to note that the last two problems were alike in structure, and the given information was the same. In the first problem, the events were dependent, but in the second problem the events were independent. The point is we must check for independence using the following definition:

Independent Events Formula

$P(A|B) = P(A)$ <u>if and only if</u> A and B are independent.

A Study Guide For Statistics

Hot Spot #4 – Sample Problem and Solution: Consider the toss of two six sided dice. Let A equal the event *the sum is even* and B equal the event *the sum is a multiple of three*. Determine if A and B are independent.

Solution: There are 36 pairs of outcomes when throwing two dice. (See the tree diagram on page 4–12 of the Study Guide).

$$P(A) = \tfrac{18}{36} \qquad P(A \text{ and } B) = \tfrac{6}{36} \qquad P(B) = \tfrac{12}{36} \qquad P(A|B) = \tfrac{6}{12}$$

Since $P(A) = \tfrac{1}{2}$ and $P(A|B) = \tfrac{1}{2}$, A and B are independent.

Sample Problem and Answer: Given the toss of one six sided die, let A equal the event *the number is more than three* and let B equal the event *the number is a multiple of three*. Determine if A and B are independent.

Answer: $P(A) = \tfrac{3}{6} = \tfrac{1}{2}$ and $P(A|B) = \tfrac{1}{2}$. Therefore, A and B are independent.

Sample Problem and Answer: Consider the toss of a six sided die. Let A equal the event *the number is even* and B equal the event *the number is odd*. Check for independence.

Answer: $P(A) = \tfrac{3}{6}$ $P(A|B) = \tfrac{0}{6}$. Therefore, A and B are dependent.

Hot Spot #5 – Mutually Exclusive Versus Independent Events

When A and B cannot both occur together, we say the two events are mutually exclusive. Another way of looking at the idea is that the first event excludes the second event. This is illustrated in tossing a die and letting A = odd number and B = even number with $S = \{1,2,3,4,5,6\}$. Events A and B are said to be mutually exclusive. Whenever we have $P(A|B) = 0$ the two events are said to be mutually exclusive:

$$P(A|B) = 0 \text{ because } P(A \text{ and } B) = \frac{n(A \text{ and } B)}{n(S)} \text{ and } n(A \text{ and } B) = 0.$$

It is important to understand the difference between events that are mutually exclusive and those that are independent. A simple way of relating the two ideas is that if two events are mutually exclusive they must be dependent; and therefore not independent. For example, consider two events A and B that are mutually exclusive, with $P(A) = 0.5$:

$P(A \text{ and } B) = 0$ because A and B are mutually exclusive.

Using the formula for $P(A|B)$, we get

$$P(A|B) = \frac{P(A \text{ and } B)}{P(B)} = \frac{0}{P(B)} = 0.$$

Since $P(A) = 0.5$ and $P(A|B) = 0$, the events are dependent.

We must be careful not to get careless in our thinking. It may seem that mutually exclusive events and dependent events are one and the same. This is clearly **not** the case. The following example involves two events that are dependent but not mutually exclusive. The problem of finding the probability of getting a 3 given a multiple of 3 on the toss of one die is done as follows:

Let $A = \{3\}$ $B = \{3,6\}$ with $S = \{1,2,3,4,5,6\}$

$$P(A|B) = \frac{P(A \text{ and } B)}{P(B)} = \frac{\frac{1}{6}}{\frac{2}{6}} = \frac{1}{2} \qquad P(A) = \frac{1}{6}$$

Since $P(A|B) \neq P(A)$, A and B are dependent.

Since $P(A \text{ and } B) \neq 0$, A and B are **not** mutually exclusive.

Hot Spot #6 – Special Vs. General Formula For $P(A \text{ and } B)$ & $P(A \text{ or } B)$

✳
Mann
Sections
4.8.2 And 4.9.2

There are two formulas in the chapter that we have not used so far in this review. They are referenced as the addition and the multiplication rules. Actually, we have used the formulas informally as we have worked through some of the problems.

When we considered the probability of events A or B, we simply added the probabilities when the events were mutually exclusive. But not all events are mutually exclusive; so we need what is called the addition rule that is used when we want the probability of A or B.

$$P(A \text{ or } B) = P(A) + P(B) - P(A \text{ and } B)$$

P(A or B)
General Case

In a similar way, when we worked with Tree Diagrams and considered the probability of A and B, we multiplied the probabilities together. If we do not use a Tree Diagram, it may not be clear whether one event is influencing another; as it happened when we drew cards without replacement. As a result, we need what is called the multiplication formula that is used when we want the joint probability of A and B.

$$P(A \text{ and } B) = P(A)P(B|A)$$

or equivalently

$$P(A \text{ and } B) = P(B)P(A|B)$$

P(A and B)
General Case

In the special cases of independent events and mutually exclusive events, the addition and multiplication formulas can be written in much simpler forms.

If events A and B are mutually exclusive, the joint probability $P(A \text{ and } B) = 0$. The addition rule then becomes the sum of the marginal probabilities:

P(A or B)
Special Case

$$P(A \text{ or } B) = P(A) + P(B)$$

If events A and B are independent, $P(B|A) = P(B)$, then the multiplication rule for the joint probability of A and B becomes the product of the marginal probabilities:

P(A and B)
Special Case

$$P(A \text{ and } B) = P(A)P(B)$$

Clearly, it is an advantage to be aware of the special cases. But as a rule, it is better to learn the more general formulas that will work for any events, whether or not they are mutually exclusive or independent.

✳

*Mann
Sections
4.4, 4.5, 4.6, 4.7,
4.8, and 4.9*

Hot Spot #7 – Deciding On Which Formula To Use

One of the most difficult things to do in this chapter is to decide on which formula to use. There are a few keys that will help us decide on the correct formula. For example, the use of the phrase "probability of at least one" suggests the complement formula $P(\overline{A}) = 1 - P(A)$. The phrase "determine if events are independent" tells us to use

$$P(A|B) = \frac{P(A \text{ and } B)}{P(B)}.$$

A Study Guide For Statistics

Another way to help us decide on the correct formula is to become familiar with the formulas. A good way to do this is to do problems that require we know and correctly use appropriate formulas. The following example illustrates the idea:

Given $P(A) = 0.5$, $P(A \text{ and } B) = 0.25$; find $P(B|A)$.

Step 1. We need a formula that contains $P(B|A)$ so we use
$P(A \text{ and } B) = P(A)P(B|A)$.

Step 2. We substitute the given information into the formula, getting $0.25 = (0.5)P(B|A)$.

Step 3. We then solve for $P(B|A) = \dfrac{0.25}{0.5} = 0.5$

Hot Spot #7 Sample Problem and Solution: Given $P(A \text{ and } B) = 0.25$, $P(A) = 0.5$, $P(A \text{ or } B) = 0.7$; find $P(B)$.

Solution: Step 1. We need a formula that includes $P(A)$, $P(A \text{ and } B)$. $P(A \text{ or } B)$ and $P(B)$, so we use

$P(A \text{ or } B) = P(A) + P(B) - P(A \text{ and } B)$.

Step 2. We substitute getting $0.7 = 0.5 + P(B) - 0.25$.

Step 3. We then solve for $P(B) = 0.7 - 0.5 + 0.25 = 0.45$.

Sample Problem and Answer: Given $P(\overline{A}) = 0.3$; find $P(A)$.

Answer: Step 1. We need the formula for the complement
$$P(A) = 1 - P(\overline{A}).$$

Step 2. We substitute, getting $P(A) = 1 - 0.3 = 0.7$.

Sample Problem and Answer: Given $P(A|B) = 0.2$, $P(\overline{B}) = 0.3$, find $P(A \text{ and } B)$.

Answer: Step 1. We need two formulas: $P(B) = 1 - P(\overline{B})$ and
$$P(A|B) = \frac{P(A \text{ and } B)}{P(B)}.$$

Step 2. We use the first formula and solve for $P(B)$.
$$P(B) = 1 - 0.3 = 0.7$$

Step 3. We then use the second formula and substitute the value for $P(B)$ and $P(A|B)$.
$$0.2 = \frac{P(A \text{ and } B)}{0.7}$$

Step 4. We solve.
$$P(A \text{ and } B) = (0.2)(0.7) = 0.14$$

Hot Spot #8 – $P(A \text{ and } B)$ **Versus** $P(A|B)$

Mann
Section 4.8.2

There is one particularly confusing situation involving the multiplication formula. The formula can be written in two ways as follows:

$$P(A \text{ and } B) = P(A)P(B|A) \text{ or } P(B|A) = \frac{P(A \text{ and } B)}{P(A)}.$$

In the first form we are looking at the joint probability of A and B, and we call it the multiplication rule. But in the second form, we are focusing on the probability of A given B. We get the second form by using the first formula and solving for $P(B|A)$. The second form is referred to as the formula for conditional probability.

The confusion occurs when we are asked to find $P(B|A)$, and we realize we need $P(A \text{ and } B)$. It seems like we should be able to use the first formula to find $P(A \text{ and } B)$. But we then realize that this requires $P(B|A)$, which is what we wanted in the beginning. We have just gone full-circle; like a puppy chasing its tail. We are not able to do what we wanted to do. Instead, we must go back to the sample space to find the $P(A \text{ and } B)$. We then substitute our value into the equation

$$P(B|A) = \frac{P(A \text{ and } B)}{P(A)}$$

to get our answer.

Another problem occurs when we realize that there are really two variations of the formula for what we call conditional probability. We can write either

Conditional
Probability
Formula

$$P(A|B) = \frac{P(A \text{ and } B)}{P(B)} \text{ or } P(B|A) = \frac{P(A \text{ and } B)}{P(A)}.$$

The situation is not so nearly confusing if we note the following:

- In the phrase $P(A|B)$, the $A|B$ reminds us of the fraction
$$\frac{A}{B},$$
whose denominator is B.

- On the right side of the equal sign we have
$$\frac{P(A \text{ and } B)}{P(B)},$$
whose denominator involves the same letter B.

We can restate the pattern as follows:

The letter in the denominator on the right side of the equal sign is the same as the letter that follows the vertical bar (|) on the left side of the formula.

Using A Pattern
To Find The
Denominator In
The Formula For
Conditional
Probability

$$P(B|A) = \frac{P(A \text{ and } B)}{P(A)} \longleftarrow \text{letter in denominator on right side}$$

$\uparrow$

letter after vertical bar

CHAPTER 4 DISCUSSION QUESTIONS

These questions may be used in your study group or simply as topics for individual reflection. Whichever you do, take time to explain verbally each topic to insure your own understanding. Since these questions are intended as topics for discussion, answers to these questions are not provided. If you find that you are not comfortable with either your answers or that your group has difficulty with the topic, take time to meet with your professor to get help.

1. What is the difference between dependent and independent events?

2. Why is the probability of A and B equal to zero when A and B are mutually exclusive?

3. When using a Tree Diagram, why is the number of branches at the end the same as the number of total possible outcomes?

4. What is the difference between marginal probability and conditional probability?

5. What is the difference between the relative frequency concept and the classic definition of probability?

6. When does one use the probability formula $P(A) = 1 - P(\overline{A})$?

7. When is the formula $P(A \text{ or } B) = P(A) + P(B)$ used?

8. When is the formula $P(A \text{ and } B) = P(A)P(B)$ used?

9. When drawing from a container without replacement, what form of the formula for the probability of A and B is used?

10. What rule is used to help decide the number of possible outcomes for a probability experiment?

CHAPTER 4 TEST

1.   Draw the Tree Diagram for the experiment of tossing a four sided die (Tetrahedron) where a 1, 2, 3, and 4 are possible. Indicate outcomes only.

2.   Fill in the appropriate probabilities on the following Tree Diagram for answering a five part multiple choice test question when guessing.

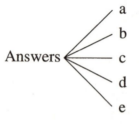

3.   How many possible outcomes are there when both a coin and a four sided die are tossed?

4.   If $P(\overline{B}) = 0.7$ $P(A|B) = 0.1$, find $P(A \text{ and } B)$.

5.   Determine if the following events involving the throw of a six sided die are independent:

$A =$ getting a multiple of 2 $B =$ getting a 4 or 6

6.   Consider the experiment of drawing two cards in succession without replacement; drawing a face card is a success.
Find the probability of $B =$ drawing a face card on the second draw given $A =$ a face card was drawn on the first draw.

7.   Find the probability of at least one head in four tosses of a fair coin (heads and tails have the same chance of occurring). Document your answer using the appropriate formula.

8. If events A and B are independent and $P(A) = 0.25$ and $P(B) = 0.75$, find $P(A$ and $B)$ and use your answer to find $P(A$ or $B)$. Again document your answers by using the appropriate formulas.

9. Find the $P(A|B)$ where A = multiple of 5 and B = sum greater than 5 (tossing two dice).

10. Give a counter example for the following statement: If two events are dependent they must be mutually exclusive.

CHAPTER 4 TEST Questions and Answers

1. Draw the Tree Diagram for the experiment of tossing a four sided die (Tetrahedron) where a 1, 2, 3, and 4 are possible. Indicate outcomes only.

Answer:

Tree Diagram

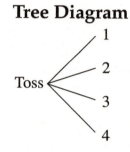

2. Fill in the appropriate probabilities on the following Tree Diagram for answering a five part multiple choice test question when guessing.

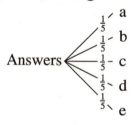

Answer:

Tree Diagram

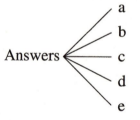

3. How many possible outcomes are there when both a coin and a four sided die are tossed?

Answer: Total possible outcomes $= 2 \times 4 = 8$

4. If $P(\overline{B}) = 0.7$ $P(A|B) = 0.1,$ find $P(A \text{ and } B)$.

Answer: $P(B) = 1 - P(\overline{B}) = 1 - 0.7 = 0.3$
$P(A \text{ and } B) = P(B)P(A|B) = (0.3)(0.1) = 0.03.$

5. Determine if the following events involving the throw of a six-sided die are independent:

$A =$ getting a multiple of 2 $B =$ getting a 4 or 6

Answer: $P(A) = \dfrac{3}{6}$ $P(A|B) = \dfrac{P(A \text{ and } B)}{P(B)} = \dfrac{\frac{2}{6}}{\frac{2}{6}} = 1$

Since $P(A) \neq P(A|B)$, the events A and B are **not** independent.

6. Consider the experiment of drawing two cards in succession without replacement; drawing a face card is a success.
Find the probability of $B =$ drawing a face card on the second draw given $A =$ a face card was drawn on the first draw.

Answer: $P(B|A) = \dfrac{11}{51}$

7. Find the probability of at least one head in four tosses of a fair coin (heads and tails have the same chance of occurring). Document your answer using the appropriate formula.

Answer: $P(\text{at least one head}) = 1 - P(\text{no heads})$

$$= 1 - \left(\tfrac{1}{2}\right)^4 = 1 - 0.0625 = 0.9375$$

8. If events A and B are independent and $P(A) = 0.25$ and $P(B) = 0.75$, find $P(A \text{ and } B)$ and use your answer to find $P(A \text{ or } B)$. Again document your answers by using the appropriate formulas.

Answer: $P(A \text{ and } B) = P(A)P(B) = (0.25)(0.75) = 0.1875$

$\qquad P(A \text{ or } B) = P(A) + P(B) - P(A \text{ and } B)$

$\qquad\qquad = 0.25 + 0.75 - 0.1875 = 0.8125$

9. Find the $P(A|B)$ where $A = $ multiple of 5 and $B = $ sum greater than 5 (tossing two dice).

Answer: $P(A|B) = \dfrac{P(A \text{ and } B)}{P(B)} = \dfrac{\frac{3}{36}}{\frac{26}{36}} = \dfrac{3}{26}$

10. Give a counter example for the following statement: If two events are dependent they must be mutually exclusive.

Answer: Test question #5 provides the counter example.

A and B are dependent because $P(A) \neq P(A|B)$

A and B are **not** mutually exclusive because $P(A \text{ and } B) \neq 0$

Therefore we have dependent events that are **not** mutually exclusive.

$\Sigma\circ$

chapter five

DISCRETE RANDOM VARIABLES AND THEIR PROBABILITY DISTRIBUTIONS

They are called wise
who put things in their right order.

– St. Thomas Aquinas

INTRODUCTION

In this chapter, we bring together many of the ideas developed in the first four chapters. We will again work with means and standard deviations and the shapes of distributions. But this time, we will be working with values from the population itself. The populations will be defined using ideas we first saw in the chapter on probability.

We start with the idea of a **random variable** that assigns a number to the outcomes of an experiment. The random variable is classified as either a **continuous random variable** or a **discrete random variable**. The probabilities related to the outcomes are then combined with the random variable to create a **probability distribution**.

Specifically, we will work with **discrete probability distributions** in this chapter. This is the first part of our work with probability distributions that will continue into the next chapter where we study continuous probability distributions that will be used to do what we will call a hypothesis test in inferential statistics.

In the earlier chapters we looked at collections of sample data. We then calculated sample statistics as estimates of the population parameters, and we were concerned that we had representative samples. Now we have an opportunity to look at the population itself through the probability distribution. Given a probability distribution, it will now be possible to find the **population mean**, and the **population standard deviation**. A very useful concept called the **expected value** of a random variable will be defined as the population mean.

The graphs that we created in the earlier chapters provided us an idea about the shape of the distribution of the population. In this chapter we create a **probability histogram** using bars to represent the probability for each value of the random variable. The probability histogram, like the grouped frequency distribution histogram, provides insight into the shape of the population defined by the probability distribution.

We will look at three important, and useful, discrete probability distributions. The first is named the **binomial probability distribution**. It involves a **binomial experiment** with repeated independent trials resulting in one of two possible outcomes, success or failure. When the trials are not independent, we use the **hypergeometric probability distribution**. The third, the **Poisson probability distribution**, involves independent occurrences of an event in a fixed interval. These three distributions are involved in a large number of applications.

This chapter is important to what follows in that we learn how to find the probability of getting specific values of the random variable. In particular, we relate the probabilities to the area in the bars of the probability histogram. We are interested in events that are rare; events that have less than a 5% chance of occurring. When such events do occur, we consider them significant. These ideas will be used in hypothesis testing in later chapters.

A Study Guide For Statistics

CHAPTER 5 HOT SPOTS

If you find other HOT SPOTS, write them down and use them as a focus of your discussions in the study group. Or you can use the HOT SPOT as the topic for a help session with your professor.

USE THIS PAGE FOR KEEPING TRACK OF TERMS AND NOTATION

Hot Spot #1 – Discrete Versus Continuous Random Variables

✳
MANN
Sections
5.1.1 And 5.1.2

The definitions for both discrete and continuous random variables are rather difficult to understand at first. There is a fairly simple way of thinking about discrete and continuous random variables. If the variable represents something that is counted, then the random variable is discrete. If the variable represents something that is measured, then the random variable is continuous.

It is not enough to just look at the values given for the random variable. For example, if the random variable represents the time to do a statistics problem on a test, we might simply give the time to the nearest five minutes. Even though the time is reported as an integer, the measurement of time could have been any value in an interval of 0 to 60 minutes. Hence, the random variable in this case would be continuous.

It is much easier to determine when we are counting. For example, in this chapter we will count the number of successes in the repeated trials of an experiment. The result is a discrete random variable.

Again, to determine whether the random variable is discrete or continuous, it is best to simply ask if we counted or measured to get the value.

$$\text{COUNT} \longrightarrow \text{DISCRETE}$$

$$\text{MEASURE} \longrightarrow \text{CONTINUOUS}$$

See the problem set in sections 5.1.1 and 5.1.2 for examples of both discrete and continuous variables.

Hot Spot #2 – Checking For A Probability Distribution

It is easy to check if we have a probability distribution. We must satisfy the following two conditions:

$$0 \le P(x) \le 1 \text{ for all } x$$

$$\sum P(x) = 1$$

The first condition states that the probability of x must be between 0 and 1 for all values of the random variable. The second condition states that the sum of the probabilities for all possible values of the random variable must equal 1. The following example illustrates what should be done to check on a potential probability distribution. Determine if the following function is a probability distribution:

$$P(x) = \frac{x}{5} \text{ for } x = 1, 2, 3, 4, 5$$

We first check to see if $P(x)$ is between 0 and 1 for all x.

$$P(1) = \tfrac{1}{5}$$
$$P(2) = \tfrac{2}{5}$$
$$P(3) = \tfrac{3}{5}$$
$$P(4) = \tfrac{4}{5}$$
$$P(5) = \tfrac{5}{5}$$

The function satisfies the first check. Now we need to find the sum of all the probabilities.

$$P(1) + P(2) + P(3) + P(4) + P(5) = \tfrac{1}{5} + \tfrac{2}{5} + \tfrac{3}{5} + \tfrac{4}{5} + \tfrac{5}{5}$$

$$\sum P(x) = 3$$

The function fails the second check.

Hot Spot #2 Sample Problem and Solution: Determine if the following function is a probability distribution function:

$$P(x) = \frac{x^2}{3} \text{ for } x = 1,2,3$$

Solution: $P(x) > 1$ for $x = 2$ and $x = 3$.

      This is not a probability distribution function.

Sample Problem and Answer: Determine if the following table is a probability distribution:

x	$P(x)$
0	0
1	0.1
2	0.4
3	0.9

Answer: The sum of the probabilities is more than 1.

      This is not a probability distribution.

Sample Problem and Answer: Determine if the following is a probability distribution:

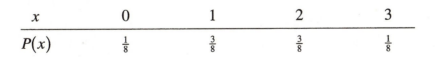

x	0	1	2	3
$P(x)$	$\frac{1}{8}$	$\frac{3}{8}$	$\frac{3}{8}$	$\frac{1}{8}$

Answer: Each probability is between 0 and 1, and the sum of the probabilities is 1; therefore, this is a probability distribution.

Hot Spot #3 – Determining If We Have A Binomial Random Variable

Deciding if the random variable is a binomial random variable at first seems like one of the more difficult things to do in this chapter. However, there are only three requirements we need to check.

The first requirement is that the problem involves repeated trials. This is simply a statement that the same thing is done each time. The problems often involve tossing a die repeatedly or flipping a coin repeatedly; but can involve any repetitious task such as doing test questions, shooting baskets, etc.

The second requirement is that each trial is identical and results in one of two possible outcomes. One of the outcomes will be considered a success, the other a failure. The success must be the same in each trial. For example, if we toss a die and decide that getting a 3 is a success, we can not change our mind after 5 tosses and decide that getting a 6 is a success.

The third requirement is that the probability must be the same for each trial. This is not a problem when we toss dice or flip coins. It may indeed be a problem when we shoot baskets or do test questions. In such cases, we may need to make an assumption that the probability of a success does not change from trial to trial.

Hot Spot #4 – Using $P(x) = \binom{n}{x} p^x q^{n-x}$ For Binomial Random Variables

There is an important pattern to observe when using the binomial formula to calculate probabilities for a binomial random variable. The exponent x used on p, the probability of success, is the same x in $P(x)$. For example, if $n = 5$ and $p = 0.2$, then $P(3)$ is written as follows:

$$P(3) = \binom{5}{3}(0.2)^3(0.8)^2$$

We need to notice where the 3's are placed. We have the first 3 in $P(3)$. The second 3 is in $\binom{5}{3}$. The last 3 is in $(0.2)^3$. This pattern is always used when we use the binomial probability formula to find $P(x)$.

Hot Spot #4 Sample Problem and Solution: Given x has a binomial distribution with $n = 4$ and $p = 0.3$, write the expression for $P(3)$.

Solution: Step 1. The 3 in $P(3)$ is the key number. We write $\binom{n}{x}$ as $\binom{4}{3}$

and p^x as p^3.

Step 2. The answer will be in the form $\binom{4}{3}(0.3)^3(0.7)^1$

Sample Problem and Answer: Given the binomial random variable with $n = 8$ and $p = 0.4$, write the expression for $P(5)$.

Answer: The answer will be in the form $\binom{8}{5}(0.4)^5(0.6)^3$

Sample Problem and Answer: Given the binomial distribution with $n = 5$ and $p = \frac{1}{3}$, write the expression for $P(4)$.

Answer: The answer will be in the form $\binom{5}{4}\left(\frac{1}{3}\right)^4\left(\frac{2}{3}\right)^1$

Hot Spot #5 – Computing $\begin{pmatrix} n \\ x \end{pmatrix} = \dfrac{n!}{x!(n-x)!}$ **Using Pascal's Triangle**

The expression $\dfrac{n!}{x!(n-x)!}$ is used in the formula for $P(x)$ for a binomial random variable. Another way of writing the expression is as what we call a combinatorial. We use the notation $\begin{pmatrix} n \\ x \end{pmatrix}$ to represent $\dfrac{n!}{x!(n-x)!}$. The notation $\begin{pmatrix} n \\ x \end{pmatrix}$ is used when we want to know how many different combinations are possible when we take n items x at a time.

For example, if we have 5 items such as the letters a,b,c,d and e, and want to take 3 letters at a time, we would write $\begin{pmatrix} 5 \\ 3 \end{pmatrix}$. If we only need to do one computation, the formula using factorials is not that difficult to use. But if we need to do several calculations, as would be necessary in creating a binomial probability distribution table, there is an easier way to find the values of $\begin{pmatrix} n \\ x \end{pmatrix}$.

The computation can be made very easy if we use Pascal's triangle. The triangle is created by looking at the coefficients of the binomial expansion $(p+q)^n$ for $n = 0,1,2,3,4,$, etc. The first four expansions are as follows:

$$(p+q)^0 = 1$$
$$(p+q)^1 = 1p^1 + 1q^1$$
$$(p+q)^2 = 1p^2 + 2p^1q^1 + 1q^2$$
$$(p+q)^3 = 1p^3 + 3p^2q^1 + 3p^1q^2 + 1q^3$$

We create Pascal's triangle by writing the coefficients of the terms in the following form:

$$
\begin{array}{ccccccc}
 & & & 1 & & & \\
 & & 1 & & 1 & & \\
 & 1 & & 2 & & 1 & \\
1 & & 3 & & 3 & & 1 \\
\end{array}
$$

Figure 5–1:
Pascal's Triangle

We get the next row of the triangle by adding the numbers in the row with 1 3 3 1 as follows:

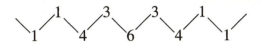

This new row gives the coefficients of the terms in $(p+q)^4$:

$$(p+q)^4 = 1p^4 + 4p^3q^1 + 6p^2q^2 + 4p^1q^3 + 1q^4$$

We get the next row of the triangle by adding the numbers in the row with 1 4 6 4 1 as follows:

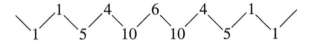

This new row gives the coefficients of the terms in $(p+q)^5$:

$$(p+q)^5 = 1p^5 + 5p^4q^1 + 10p^3q^2 + 10p^2q^3 + 5p^1q^4 + 1q^5$$

We can continue expanding the triangle to as many rows as we need.

It is important to study the pattern of the exponents in the expansions of $(p+q)^n$. We first need to observe that the sum of the exponents for each term is always equal to the value of n. Next, as we examine the terms going from left to right, the exponent on p goes down 1 each time as the exponent on q goes up 1 each time. Again, we will need to focus on the value of x in the formula $P(x) = \binom{n}{x} p^x q^{n-x}$.

Now we can use Pascal's Triangle to find $P(x)$ for a binomial random variable. The only difficulty is to decide which term of the triangle to use. All we need to do is be aware of the pattern in the terms of the expansion. If we need to find $P(x)$, the probability of x for a particular problem, we will know n and get x from the notation $P(x)$.

The value of n tells us to use the $n+1$ row of the triangle. The value of x gives us the exponent we will use on p for the appropriate term of the expansion. The following problems illustrate the process.

Hot Spot #5 Sample Problem and Solution: Given the binomial distribution with $n=3$ and $p=0.1$, find $P(2)$.

Solution: Step 1. $n=3$, so we want the row of the triangle for $(p+q)^3$.

Step 2. $P(2)$ tells us $x=2$, so we want the term $\binom{3}{2}p^2q^1$.

Step 3. Substituting for p and q we get $P(2) = \binom{3}{2}(0.1)^2(0.9)^1$.

Step 4. We go to Pascal's triangle to the row 1 3 3 1 and use the first 3 for the value of $\binom{3}{2}$.

Step 5. We can either leave the answer in the form

$P(2) = 3(0.1)^2(0.9)^1$, or perform the computation to get

$P(2) = 0.027$.

In Mann The Answers Are Written In The Decimal Form.

Sample Problem and Answer: Given the binomial distribution with $n=4$ and $p=0.3$, find $P(1)$.

Answer: Step 1. $n=4$, so we want the row of the triangle for $(p+q)^4$.

Step 2. $P(1)$ tells us $x=1$, so we want the term $\binom{4}{1}p^1q^3$.

Step 3. Substituting for p and q we get $P(1) = \binom{4}{1}(0.3)^1(0.7)^3$.

Step 4. We go to Pascal's triangle to the row 1 4 6 4 1 and use the second 4 for the value of $\binom{4}{1}$.

Step 5. We can either leave the answer in the form $P(1) = 4(0.3)^1(0.7)^3$ or perform the computation to get $P(1) = 0.4116$.

Sample Problem and Answer: Given the binomial random variable with $n = 5$ and $p = \frac{1}{4}$, use Pascal's triangle to find $P(x)$ for $x = 5, 4, 3, 2, 1, 0$.

Answer:

x	$P(x)$
5	$1\left(\frac{1}{4}\right)^5 = 0.0010$
4	$5\left(\frac{1}{4}\right)^4\left(\frac{3}{4}\right)^1 = 0.0146$
3	$10\left(\frac{1}{4}\right)^3\left(\frac{3}{4}\right)^2 = 0.0879$
2	$10\left(\frac{1}{4}\right)^2\left(\frac{3}{4}\right)^3 = 0.2637$
1	$5\left(\frac{1}{4}\right)^1\left(\frac{3}{4}\right)^4 = 0.3955$
0	$1\left(\frac{3}{4}\right)^5 = 0.2373$

Hot Spot #6 – When To Use $\mu = np$ And $\sigma^2 = npq$

✳
Mann
Section 5.6.5

The general formulas for the mean and variance of a discrete probability distribution are:

$$\mu = \sum xP(x) \text{ and } \sigma^2 = \sum (x - \mu)^2 P(x)$$

The formulas will work for any discrete probability distribution, but they are time consuming to use. Fortunately, we can use an easier and faster formula when we have a binomial random variable.

The formulas for the mean and variance of a binomial random variable are:

$$\mu = np \text{ and } \sigma^2 = npq$$

Remember, these shortcut formulas can only be used if the random variable is from a binomial experiment. If the formulas are used for a non-binomial discrete probability distribution, we will get an incorrect answer. The following example shows what happens when we use the wrong formula.

The following function is a probability distribution:

x	1	2	3	4	5	6
$P(x)$	$\frac{1}{6}$	$\frac{1}{6}$	$\frac{1}{6}$	$\frac{1}{6}$	$\frac{1}{6}$	$\frac{1}{6}$

We can use the formula $\mu = \sum xP(x)$ to find the mean because this formula will give us the mean for any discrete probability distribution.

$$\mu = 1\left(\tfrac{1}{6}\right) + 2\left(\tfrac{1}{6}\right) + 3\left(\tfrac{1}{6}\right) + 4\left(\tfrac{1}{6}\right) + 5\left(\tfrac{1}{6}\right) + 6\left(\tfrac{1}{6}\right)$$

$$\mu = (1 + 2 + 3 + 4 + 5 + 6)\left(\tfrac{1}{6}\right) = \tfrac{21}{6} = 3.5$$

Now, we will make the faulty assumption that we have a binomial distribution and use the wrong formula for the problem.

$$\mu = np = 6\left(\tfrac{1}{6}\right) = 1$$

In this problem, it is very clear that the distribution is not a binomial. Furthermore, it is easy to see that the distribution is not centered at $\mu = 1$. Hopefully, we would see our mistake if we used the wrong formula. Of course, it is best to avoid the mistake in the first place and use the correct formula. When in doubt, do not use the shortcut formula.

Hot Spot #7 – Interpreting The Probability Of At Least k Successes

✳
*Mann
Sections
5.6.3 And 5.7.1*

The phrase, "find the probability of at least k successes", often causes some difficulty. Part of the problem is to understand what the phrase, *at least,* means. For example, *at least* 2 means 2 or more. So if $n = 5$ in a discrete probability distribution, the phrase *at least* 2 would mean that we are interested in 2, 3, 4, or 5 successes.

The probability of at least 2 successes would be written as $P(x \geq 2)$ and include $P(x = 2) + P(x = 3) + P(x = 4) + P(x = 5)$. The idea of a complement from the probability chapter could be used to simplify the problem. We would write $P(x \geq 2)$ as $1 - P(x \leq 1)$. Then, we would only have to find $P(x = 0) + P(x = 1)$ and subtract the sum from 1.

Hot Spot #7 Sample Problem and Solution: Given the binomial distribution with $n = 5$ and $p = 0.4$, find the probability of at least one success.

Solution: Step 1. We want to find $P(x \geq 1)$. We use the complement to write the following:

$$P(x \geq 1) = 1 - P(x < 1) = 1 - P(x = 0)$$

Step 2. $P(x = 0) = C_{5,0}(0.4)^0(0.6)^5 = 0.07776$

Step 3. $P(x \geq 1) = 1 - 0.07776 = 0.92224$

Sample Problem and Answer: Given the binomial distribution with $n = 4$ and $p = 0.2$, find the probability of at least 2 successes.

Answer: $P(x \geq 2) = 1 - P(x \leq 1) = 1 - \left[P(x = 1) + P(x = 0) \right]$

$$= 1 - \left[4(0.2)^1 (0.8)^3 + 1(0.8)^4 \right]$$

$$= 1 - (0.4096 + 0.4096) = 1 - 0.8192 = 0.1808$$

Sample Problem and Answer: Given the Poisson distribution with $n = 20$ and $\lambda = 0.5$, find the probability of at least one success.

Answer: $P(x \geq 1) = 1 - P(x \leq 0) = 1 - \dfrac{(0.5)^0 e^{-.5}}{0!} = 1 - 0.6065 = 0.3935$

Hot Spot #8 – Binomial Probability Distribution Tables

Mann
Section 5.6.3

When we work with a binomial random variable, there are two common options that are used to find the probability for a specific value of the random variable. The easiest method is to use either a computer or a calculator. When we input a value of n and p, the calculator or computer will give us the probability we want.

The second method is to use a table of probabilities. Binomial probability tables are readily available for small values of n and common values of p, such as 0.1, 0.2, 0.3, etc. Later on we will work with a method that gives approximate answers for large values of n. However, there may still be a need to construct tables for value of p not found in a typical set of tables. We can then use the binomial probability formula $P(x) = \dbinom{n}{x} p^x q^{n-x}$ to construct a table of binomial probabilities. The easiest way to do the problem is using the appropriate expansion of $(p + q)^n$ as shown in the following sample problems.

Hot Spot #8 Sample Problem and Solution: Create the probability table for the binomial probability distribution with $n = 3$ and $p = 0.4$.

Solution: Step 1.   We want $(p + q)^3 = 1p^3 + 3p^2q^1 + 3p^1q^2 + 1q^3$.

Step 2.   We create the table as follows:

See Hot Spot #5 for help with using Pascal's Triangle.

x	$P(x)$
3	$1p^3$
2	$3p^2q^1$
1	$3p^1q^2$
0	$1q^3$

Step 3.   We now substitute for p and q as follows:

x	$P(x)$
3	$1(0.4)^3$
2	$3(0.4)^2(0.6)^1$
1	$3(0.4)^1(0.6)^2$
0	$1(0.6)^3$

Step 4.   We can write the values with decimals as follows:

x	$P(x)$
3	0.064
2	0.288
1	0.432
0	0.216

Sample Problem and Answer: Given the binomial random variable with $n = 2$ and $p = 0.8$, create the binomial probability table.

Answer:

x	$P(x)$
2	$1(0.8)^2$
1	$2(0.8)^1(0.2)^1$
0	$1(0.2)^2$

Sample Problem and Answer: Given the binomial distribution with $n = 5$ and $p = \frac{1}{3}$, create the table of binomial probabilities.

Answer:

x	$P(x)$
5	$1\left(\frac{1}{3}\right)^5$
4	$5\left(\frac{1}{3}\right)^4\left(\frac{2}{3}\right)^1$
3	$10\left(\frac{1}{3}\right)^3\left(\frac{2}{3}\right)^2$
2	$10\left(\frac{1}{3}\right)^2\left(\frac{2}{3}\right)^3$
1	$5\left(\frac{1}{3}\right)^1\left(\frac{2}{3}\right)^4$
0	$1\left(\frac{2}{3}\right)^5$

CHAPTER 5 DISCUSSION QUESTIONS

These questions may be used in your study group or simply as topics for individual reflection. Whichever you do, take time to explain verbally each topic to insure your own understanding. Since these questions are intended as topics for discussion, answers to these questions are not provided. If you find that you are not comfortable with either your answers or that your group has difficulty with the topic, take time to meet with your professor to get help.

1. What is the difference between a discrete random variable and a continuous random variable?

2. What is a random variable?

3. What is a probability distribution?

4. How do you check to see if you have a probability distribution?

5. What is the difference between a binomial random variable and a Poisson random variable?

6. What is the expected value of a random variable?

7. When can you use $\mu = np$ and $\sigma^2 = npq$?

8. What is meant by the probability of at least one?

9. Describe a binomial experiment.

10. How is Pascal's Triangle used to find the coefficients of $(p + q)^n$?

CHAPTER 5 TEST

1. Given the discrete probability distribution below, find μ:

x	0	1	2	3
$P(x)$	0.1	0.2	0.3	.0.4

2. Use the probability distribution in problem #1 and find σ^2.

3. Find $E(x)$ for the following probability distribution:

x	-2	3
$P(x)$	$\frac{1}{3}$	$\frac{2}{3}$

4. Given the binomial distribution with $n = 5$ & $p = \frac{1}{2}$, find μ and σ^2.

5. Draw the probability histogram for the binomial distribution in #4.

6. Find the value of $\dfrac{n!}{x!(n-x)!}$ for $n = 5$ and $x = 3$.

7. Given the Poisson distribution with $n = 3$ and $\lambda = 0.6$. Find the probability of at least one success.

8. Given the binomial random variable with $n = 5$ and $p = 0.2$ use the formula $P(x) = \dbinom{n}{x} p^x q^{n-x}$ to find $P(3)$.

9. Write out the probability distribution table for the binomial random variable with $n = 2$ and $p = \frac{1}{2}$.

10. Determine if the following table represents a discrete probability distribution:

x	0	1	2	3
$P(x)$	0.125	0.375	0.375	0.125

CHAPTER 5 TEST Questions and Answers

1. Given the discrete probability distribution below, find μ:

x	0	1	2	3
$P(x)$	0.1	0.2	0.3	0.4

Answer: $\mu = \sum xP(x) = 0(0.1) + 1(0.2) + 2(0.3) + 3(0.4) = 2$

2. Use the probability distribution in problem #1 and find σ^2.

Answer: $\sigma^2 = \sum x^2 P(x) - \mu^2 = 0(0.1) + 1(0.2) + 4(0.3) + 9(0.4) - 4 = 1$

3. Find $E(x)$ for the following probability distribution:

x	-2	3
$P(x)$	$\frac{1}{3}$	$\frac{2}{3}$

Answer: $E(x) = \sum xP(x) = (-2)\left(\frac{1}{3}\right) + (3)\left(\frac{2}{3}\right) = -\frac{2}{3} + \frac{6}{3} = \frac{4}{3}$

4. Given the binomial distribution with $n = 5$ & $p = \frac{1}{2}$, find μ and σ^2.

Answer: $\mu = np = (5)\left(\frac{1}{2}\right) = 2.5$ and $\sigma^2 = npq = 5\left(\frac{1}{2}\right)\left(\frac{1}{2}\right) = \frac{5}{4}$

5. Draw the probability histogram for the binomial distribution in #4.

Answer: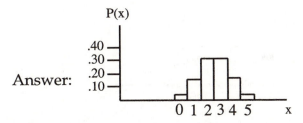

6. Find the value of $\dfrac{n!}{x!(n-x)!}$ for $n = 5$ and $x = 3$.

Answer: $\dfrac{5!}{(3!)(2!)} = \dfrac{(5)(4)(3)(2)(1)}{[(3)(2)(1)][(2)(1)]} = \dfrac{(5)(4)(3)(2)(1)}{(3)(2)(1)(2)(1)} = \dfrac{20}{2} = 10$.

7. Given the Poisson distribution with $n = 3$ and $\lambda = 0.4$. Find the probability of at least one success.

Answer: $P(x \geq 1) = 1 - P(x \leq 0) = 1 - \dfrac{(0.6)^0 e^{-.6}}{0!} = 1 - 0.5488 = 0.4512$

8. Given the binomial random variable with $n = 5$ and $p = 0.2$ use the formula $P(x) = \dbinom{n}{x} p^x q^{n-x}$ to find $P(3)$.

Answer: $P(3) = \dbinom{5}{3}(0.2)^3(0.8)^2 = 10(0.008)(0.64) = 0.0512$.

9. Write out the probability distribution table for the binomial random variable with $n = 2$ and $p = \frac{1}{2}$.

Answer:

x	$P(x)$
0	$\frac{1}{4}$
1	$\frac{1}{2}$
2	$\frac{1}{4}$

10. Determine if the following table represents a discrete probability distribution:

x	0	1	2	3
P(x)	0.125	0.375	0.375	0.125

Answer: $P(x)$ is between 0 and 1 for every value of x. The sum of $P(x)$ for all values of x equals 1. The table is a discrete probability distribution.

Σ∘

chapter six

CONTINUOUS RANDOM VARIABLES AND THE NORMAL DISTRIBUTION

Everybody believes in the [normal approximation],
the experimenters because
they think it is a mathematical theorem,
the mathematicians because
they think it is an experimental fact.

— G. Lippmann

INTRODUCTION

Many of the ideas in this chapter on **continuous probability distributions** are extensions of material we have already seen in the chapter on discrete probability distributions. The idea of probability as area will again be used. But the area will be under a curve rather than in the rectangles of the probability histograms we used in the last chapter.

There are several key differences in this chapter. In very simple terms, a discrete variable meant that we were counting something. In this chapter, a **continuous variable** is the result of measuring something rather than counting. There is a major difference in the probability statements we will make using discrete and continuous variables.

In a discrete probability distribution, it was possible to find the probability of getting a specific value of the random variable. When we work with a continuous random variable, the probability of the variable being equal to a specific value is zero. But we are able to find the probability that a variable is in an interval. The **Empirical Rule** will be used to find the probability that the continuous variable is in an interval about the mean.

This chapter provides an important transition between the idea of hypothesis testing and a probability distribution; first introduced in the last chapter and then extended to continuous variables in this chapter. Throughout this chapter, we will use the z score that gives the distance between the mean and the point represented by z in terms of the standard deviation. We will refer to z as the **standard score**; it is also known as the standard normal variable. The z score and the bell shaped **standard normal distribution** become the focal points in this chapter. The z score will be used to find the area under the **standard normal curve** that is used to find probabilities

The z score will continue to be a focus in the next several chapters. We will use the z score , in various forms, as a test statistic to do hypothesis testing in inferential statistics. The common structure of the various forms of the z score as a test statistic will help us organize the material in the chapters that follow.

In this chapter, notation continues to be very important to us as a way of organizing our work and understanding what it is that we have done. We will also do a great deal of drawing; it is important that we draw a picture of a bell-shaped curve for each problem that we do.

The last thing we will look at in this chapter is the **normal approximation** to the binomial distribution. Recent advances in calculators and the use of statistical software on personal computers make this section less important than it would have been in the past decade. But the approximation will help us better understand the relationship between the discrete and continuous random variable.

CHAPTER 6 HOT SPOTS

1. **A Difference Between The Probability For Discrete And Continuous Random Variables.**
   Starts on **page 6–4.** Mann 6.1

2. **Understanding Why The Probability Of A Continuous Random Variable Equalling A Specific Value is Zero.**
   Starts on **page 6–5.** Mann 6.1

3. **Using The Empirical Rule To Find $P(a \le x \le b)$.**
   Starts on **page 6–5.** Problems on **pages 6–6, 6–7, 6–8.** Mann 6.3

4. **Using The Z Score To Find $P(x)$.**
   Starts on **page 6–8.** Problems on **pages 6–10, 6–11.** Mann 6.4

5. **Using Graphs With The Z Score.**
   Starts on **page 6–11.** Problems on **pages 6–14, 6–15.**
   Mann 6.3, 6.4, 6.5

6. **Finding c For $P(z \ge c) = 5\%$.**
   Starts on **page 6–15.** Problems on **pages 6–17, 6–18.** Mann 6.6

7. **Normal Approximation To The Binomial Distribution.**
   Starts on **page 6–18.** Problems on **pages 6–21, 6–22.** Mann 6.7

If you find other HOT SPOTS, write them down and use them as a focus of your discussions in the study group. Or you can use the HOT SPOT as the topic for a help session with your professor.

Hot Spot #1 – A Difference Between The Probability For Discrete And Continuous Random Variables

In this chapter we work with many of the same ideas we developed and used in the chapter on discrete probability distributions. However, there is one very important difference beyond the over simplified distinction that we use a discrete random variable when we count and a continuous random variable when we measure. In the work we did with discrete probability distributions, we were able to find the probability that a random variable equals a specific value. But, when we find the probability of a continuous random variable having a specific value, we get zero. (This is discussed further in **Hot Spot #2**).

For example, if the random variable x has a binomial distribution with $n = 5$ and $p = 0.5$, we can find $P(x = 3)$. We find the answer by substituting the values of n and p into the binomial probability function

$$P(x) = \frac{n!}{x!(n-x)!}p^x(1-p)^{n-x}$$

In this example, we get

$$P(x = 3) = 10\left(\tfrac{1}{2}\right)^3\left(\tfrac{1}{2}\right)^2 = 0.3125$$
$$\text{where } 10 = \frac{5!}{3!2!}$$

If x is a continuous random variable, the probability that x equals a specific value is zero. Using notation, we would write $P(x = c) = 0$, where c is any real number.

When we work with a continuous random variable, we are <u>only</u> able to find the probability that x is in an interval. For example, we are able to ask for the probability that x is between a and b. The notation we use involves the expression $P(a \leq x \leq b)$.

We then find the area between a and b under the curve defined by the continuous probability distribution.

Hot Spot #2 – Understanding Why The Probability Of A Continuous Random Variable Equalling A Specific Value Is Zero.

✳
Mann
Section 6.1

It is sometimes difficult to understand why the probability of a specific value is zero. One way of looking at the problem is to consider a large paper bag in which we write the ten digits between 0 and 9 on slips of paper and place them in the bag. If we then draw out one value, the probability of getting one of the digits between 0 and 9 would be 0.1. Now if we consider adding slips of paper to represent the "tenths" place we could find the probability of getting a specific value to the tenths place between 0 and 9.9. For example we could find the probability of getting a value like 3.4, but now we would have one hundred slips of paper in the bag. The probability of getting one of the values in the bag is now 0.01.

We could then add more slips of paper to the bag for the "hundredths" place so that we can find the probability of getting a value like 3.44. But the probability of getting a specific value is now down to 0.001. As we continue adding slips of paper to the bag to allow us to get the probability of getting any real number between 0 and 10, we would get smaller and smaller probabilities that would get closer and closer to zero. As you realize, we would never be done adding slips of paper to our bag; and we would finally simply say that the probability of getting a specific real number in the half-open interval of 0 to 10 is zero.

Hot Spot #3 – Using The Empirical Rule to Find $P(a \le x \le b)$

✳
Mann
Section 6.3

We are usually interested in finding the probability that a continuous random variable is within some interval. For example, let x equal the speed of a car on a highway in miles per hour. If x is assumed to be normally distributed with $\mu = 55$ and $\sigma = 5$, then we could find the probability that the speed of the car is more than 65.

There are two ways for us to look at this problem: using the Empirical Rule, or using a z score; which is the topic of the next **Hot Spot**. In this **Hot Spot**, we will use the Empirical Rule. Although this will be easy to do, we must realize our answer will only be an approximation of the correct answer we obtain using the z score.

We know that 95% of the speeds will be within two standard deviations of the mean. The mean is 55 mph and two standard deviations gives us 10 mph. When we use 55 for the mean, and then add and subtract the 10, we get the values of 45 and 65. So, 95% of the speeds are in the interval (45,65).

This leaves approximately 5% of the area under the bell-shaped curve outside the interval from 45 to 65. The areas outside of the interval are called the tails of the curve. The area under the right tail of the curve gives the probability of $x \geq 65$. The area under the right tail of the curve is one-half of 5%, and we can write:

$$P(x \geq 65) = 2.5\%$$

Hot Spot #3 Sample Problem and Solution: Use the Empirical Rule to find the approximate answer for $P(50 \leq x \leq 60)$ where the continuous random variable x equals the speed on the highway with $\mu = 55$ and $\sigma = 5$.

Solution: The interval between 50 and 60 is centered on 55. The graph for the bell-shaped curve centered on 55 follows (Figure 6–1):

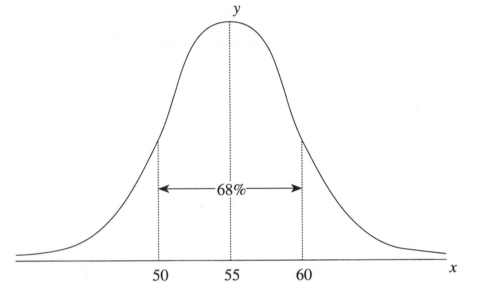

Figure 6–1:
One Standard
Deviation From
Mean of 55

The values of 50 and 60 represent the mean plus and minus one standard deviation. We know that approximately 68% of the distribution will be within one standard deviation of the mean. We can now answer the question using the appropriate notation:

$$P(50 \leq x \leq 60) = 0.68$$

Sample Problem and Answer: Let x be a continuous random variable with a normal distribution with $\mu = 1,200$ and $\sigma = 100$. Use the Empirical Rule to find an approximate value for $P(x \geq 1,300)$.

Answer: $P(x \geq 1,300) = 0.16$.

Sample Problem and Answer: Let x represent the scores on a statistics test given to all the statistics students at a university. Assume that x has a normal distribution with $\mu = 70$ and $\sigma = 10$. Use the Empirical Rule to find the approximate probability for $P(60 \leq x \leq 90)$.

Answer: $P(60 \leq x \leq 90) = 0.34 + 0.475 = 0.815$.

✳

Mann

Section 6.4

Hot Spot #4 – Using The Z Score To Find $P(x)$

An alternative to using the Empirical Rule to find $P(x)$ is to trade in questions about x for questions about z. The z score compares a data value's relationship to the mean in terms of the standard deviation. We do this so that we can obtain more accurate answers and make it easier to get the answers. Rather than having to figure out in our head how many standard deviations 65 mph is from 55 mph in our example about highway speeds with $\mu = 55$ and $\sigma = 5$, we just use the formula

$$z = \frac{x - \mu}{\sigma}$$

We substitute 65 for x and use the values of μ and σ given:

$$z = \frac{65 - 55}{5} = 2$$

The $z = 2$ tells us that we are two standard deviations away from the mean. Now we can replace our original question about the

$$P(x \geq 65) \text{ with } P(z \geq 2)$$

Remember, it does not matter if we use $\geq$ or just $>$ because the

$$P(x = 65) = 0 \text{ and } P(z = 2) = 0$$

Furthermore, it does not matter whether we use the x value or the z value. Either way, we want the area under the appropriate bell-shaped curve as shown in Figure 6–2:

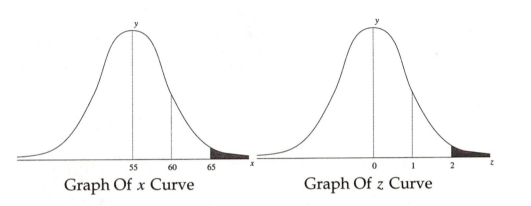

Graph Of x Curve   Graph Of z Curve

Figure 6–2:
*Graphs Of x And
z Distributions
Compared*

The real problem is to find the area under the x curve to the right of 65. There are several methods that can be used, and some methods would take a great deal of time or require a computer to get an accurate answer. For example, one of the methods we could use involves a numerical technique of integration that is done in calculus.

Fortunately, the z curve gives us an answer to the problem very quickly. All we have to do is use the z table in the back of the book. We look up the value of 2 and find that the probability of z being between 0 and 2 is 0.4772 (to the nearest ten thousandth).

Since 50% of the area under the curve is to the right of $z = 0$, we subtract our value of 0.4772 as follows:

$$P(z \geq 2) = 0.5 - 0.4772 = 0.0228$$

Figure 6–3, below, shows the bell-shaped curve for z and the area under the curve between 0 and 2. As is illustrated, we subtract the area that is shaded from 0.5 to get the area to the right of 2.

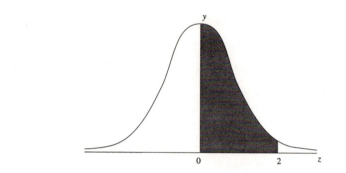

Figure 6–3: Area
Under The Curve
From 0 to 2

Compare the
next three
problems to the
corresponding
problems of
Hot Spot #3 on
pages 6–6, 6–7,
and 6–8.

Hot Spot #4 Sample Problem and Solution: Use the z value to find the exact answer for $P(50 \leq x \leq 60)$; where the continuous random variable x equals the speed on the highway with $\mu = 55$ mph and $\sigma = 5$ mph.

Solution: Step 1. Graph the bell-shaped curve centered on 55 mph.

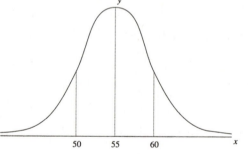

Figure 6–4:
Bell Curve
Centered At 55

Step 2. Place the values of 50 and 60 on the graph.

Step 3. We want the area under the curve between the two values. We now find the z values for each value of x.

$$z = \frac{50-55}{5} = -1 \qquad\qquad z = \frac{60-55}{5} = 1$$

Step 4. We now use the table of z values to find the area between 0 and 1.

$$P(0 \leq z \leq 1) = 0.3413$$

Step 5. Because the area between 0 and 1 is the same as the area under the curve between -1 and 0, we double the value we found in the z table. We can now answer the question using the appropriate notation:

$$P(50 \le x \le 60) = 2 \cdot P(0 \le z \le 1) = (2)(0.3413) = 0.6826$$

Sample Problem and Answer: Let x be a continuous random variable with a normal distribution with $\mu = 1,200$ and $\sigma = 100$. Use the z score to find the exact answer for $P(x \ge 1,300)$.

Answer: $P(x \ge 1,300) = 0.1587$

Sample Problem and Answer: Let x represent the scores on a statistics test given to all the statistics students at a university. Assume that x has a normal distribution with $\mu = 70$ and $\sigma = 10$. Use the z score to find the exact probability for $P(60 \le x \le 90)$.

Answer: $P(60 \le x \le 90) = 0.3413 + 0.4772 = 0.8185$

Hot Spot #5 – Using Graphs With The Z Score

✳
Mann
Sections 6.3, 6.4,
And 6.5

There is a very important point that needs to be made about working with z values. **Always take time to draw a graph!** The picture we draw helps us see the areas we need to use to get an answer. For example, if we use the normal distribution for x, where x represents highway speeds with μ=55 and $\sigma = 10$, we can find the probability that x is between 60 and 70.

We would write this using the following notation:

$$P(60 \le x \le 70)$$

The first thing we want to do is draw a diagram or graph:

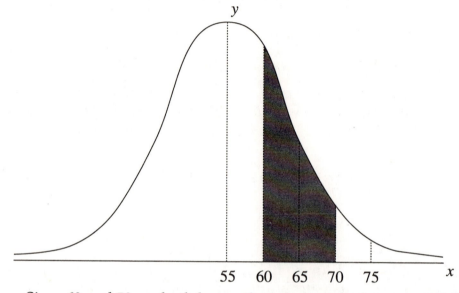

Figure 6–5:
Probability Of x
Between
60 and 70

Since 60 and 70 are both larger than the population mean of 55, the 60 and 70 are placed on the right side of the curve as shown. We want the area under the curve between 60 and 70. Using the type of approach we followed earlier, we could try to use the Empirical Rule. But this time, we find that we need the area under the bell-shaped curve between one-half and one and one-half standard deviations; $60 = 55 + 0.5(10)$ and $70 = 55 + 1.5(10)$. The Empirical Rule does not help in this situation. Unless we want to use a technique such as numerical integration from calculus, we have no choice but to use the z values that correspond to 60 and 70 as follows:

$$z = \frac{60 - 55}{10} = \frac{1}{2} = 0.5 \qquad z = \frac{70 - 55}{10} = \frac{3}{2} = 1.5$$

The graph we drew now helps us decide how to find the answer to our original question. Our question is now written as

$$P(0.5 \leq z \leq 1.5)$$

If we decide to use the z table, we will get two different areas as shown below in Figure 6–6:

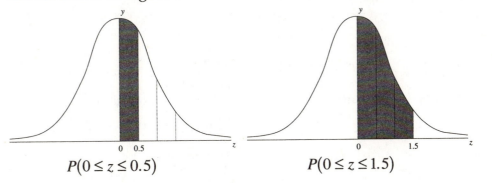

$$P(0 \leq z \leq 0.5) \qquad\qquad P(0 \leq z \leq 1.5)$$

Figure 6–6:
Two Different z
Table Areas

The larger area is the one between 0 and 1.5; we need to subtract the smaller area between 0 and 0.5 from the larger area to get our answer. We find the areas from the z table. Using the table, we look up the 1.5 value for z and find the large area is 0.4332; it is $P(0 \leq z \leq 1.5)$. When we look up 0.5 in the table, we find the small area is 0.1915; it is $P(0 \leq z \leq 0.5)$. Now, all we have to do is use the graph to show us what to do. We need to find the difference of the two areas. Even if we somehow did the subtraction backwards, we would realize our mistake because **the probability must be a positive number**. Using the values we obtained from the table we subtract as follows:

$$P(0.5 \leq z \leq 1.5) \qquad = P(0 \leq z \leq 1.5) - P(0 \leq z \leq 0.5)$$
$$= 0.4332 - 0.1915$$
$$P(0.5 \leq z \leq 1.5) \qquad = 0.2417$$

It is important to document how we get the answer using both the graph and probability notation. Many students try to take shortcuts at this point by either leaving out the notation or not drawing the graph.

Although it is possible to get the correct answer, students often get confused and make errors when they use shortcuts. So, be careful; document your work with a graph and probability notation.

Hot Spot #5 Sample Problem and Solution: Find $P(-2 \leq z \leq -0.5)$.

Solution: Step 1. Draw the bell-shaped curve.

Step 2. Locate -2 and -0.5 on the curve and shade in the area under the curve between the values.

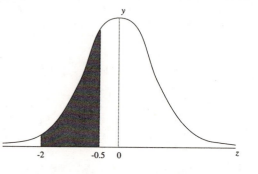

Figure 6–7:
Area Between
-2 and -0.5

Step 3. We see that we need to subtract two areas:

$$P(-2 \leq z \leq 0) - P(-0.5 \leq z \leq 0)$$

Step 4. Since the z table only uses positive values we actually find the following areas:

$$P(0 \leq z \leq 2) - P(0 \leq z \leq 0.5) \quad = 0.4772 - 0.1915$$
$$= 0.2857$$

Step 5. We should rewrite our answer as follows:

$$P(-2 \leq z \leq -0.5) \qquad = P(-2 \leq z \leq 0) - P(-0.5 \leq z \leq 0)$$
$$= 0.4772 - 0.1915 = 0.2857$$

$$P(-2 \leq z \leq -0.5) = 0.2857$$

A Study Guide For Statistics

Sample Problem and Answer: Find $P(-1.5 \leq z \leq -0.5)$.

Answer: $P(-1.5 \leq z \leq -0.5)$ $\quad = P(-1.5 \leq z \leq 0) - P(-0.5 \leq z \leq 0)$

$\quad = 0.4332 - 0.1915$

$\quad = 0.2417$

$$P(-1.5 \leq z \leq -0.5) = 0.2417$$

Sample Problem and Answer: Find $P(z \geq -1.5)$.

Answer: $P(z \geq -1.5)$ $\quad = P(-1.5 \leq z \leq 0) + 0.5$

$\quad = 0.4332 + 0.5$

$\quad = 0.9332$

$$P(z \geq -1.5) = 0.9332$$

Hot Spot #6 – Finding c For $P(z \geq c) = 5\%$

✳
Mann
Section 6.6

Related to the problem of finding the probability of z being between two values is the inverse problem of finding the value of z that gives a specific area to the right of z. For example, we may want to find the value of z that gives 5% to the right of our value. This problem can be stated in two different ways. The easiest way is to use the notation $z(0.05)$. There is a more interesting way of asking the questions as follows:

Find c so that $P(z \geq c) = 0.05$

The letter c is used to represent the actual value of z that makes the statement true. In our example we know that c must be on the right side of the curve because the area we want to the right of c is less than 50%. If we now draw the graph with c on the right side, we get the following:

Figure 6–8: Use Of c With z Scores

As is usually the case, it is very important to first draw the graph and locate c on the correct side. It is now easier to see that we need a value of c that gives 0.45 between the center of our curve and the value of c. We can use the z table to find that $c = 1.645$ gives us the 0.45 we need. We now replace the letter c with the 1.645 and write the following:

$$P(z \geq 1.645) = 0.05$$

Of all the concepts in this chapter on continuous probability distributions, this one idea seems to be very difficult to understand. Some of the difficulty seems to be with the algebra of the statement

$$z \geq c.$$

If we consider the above expression to be a problem with inequalities much like what we did in an algebra class, it may make it easier to understand what it means to have $z \geq c$. In the algebra class, we would just graph the inequality on a number line and the solution includes all of the values to the right of c as follows:

Figure 6–9: $z \geq c$ As An Inequality On A Number Line

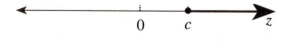

Hot Spot #6 Sample Problem and Solution: Find the value of c that makes the following statement true: $P(-c \le z \le c) = 0.95$.

Solution: Step 1. We need to locate the values of c and $-c$ on the bell-shaped curve. It is easier this time because we have an interval. The graph is drawn below in Figure 6–10:

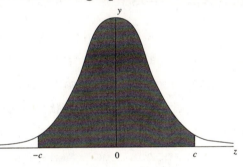

Figure 6–10:
Area Between
-c and c

Step 2. Since our values of c and $-c$ are symmetric to the center of the curve, we see that we need a value of c that will give us 0.475 between $z = 0$ and our value c. We now look up 0.475 as an area in the z table. We find a value of $c = 1.96$ gives us the required area.

Step 3. We can now write the following:
$$P(-1.96 \le z \le 1.96) = 0.95$$

Step 4. Our solution is $c = 1.96$.

Sample Problem and Answer: Find the value of c that makes the following statement true: $P(z \le c) = 0.01$.

Answer: $c = -2.326$

Sample Problem and Answer: Find the z value for $z(0.10)$.

Answer: $c = 1.282$

✳
Mann
Section 6.7

Hot Spot #7 – Normal Approximation To The Binomial Distribution

The last consideration in this chapter on continuous probability distributions involves what is called the normal approximation to the binomial distribution. The normal approximation is very useful if we do not have either a calculator or a computer that is programmed with the binomial distribution. When we have to rely on a binomial table to find probabilities, this approximation is used when either the number of trials is larger than a value listed in a table of binomial probabilities or when the p value is a value not listed in the table. The usual difficulty is illustrated in the following example.

Suppose we are asked to find the probability of getting a score of 70 or more on a 100 question True/False test and that we must assume that the individual taking the test is guessing on every question. If we let $x =$ the number correct, then x is a discrete random variable with $n = 100$ and $p = 0.5$; and we want the $P(x \geq 70)$.

The partial probability histogram below (Figure 6–11) shows the areas we want between 70 and 100; the area includes the rectangles at $x = 70$ and $x = 100$. Whenever np and nq are both greater than 5, the histogram will be approximately bell-shaped. In our example $np = nq = (100)(0.5) = 50$; so we can use the normal approximation. A bell-shaped curve has been drawn over the probability histogram. The area under the curve will be approximately equal to the area in the rectangles, and we will use the normal curve to approximate the areas in the probability histogram. The curve will be centered at $\mu = np$, the mean of the binomial distribution.

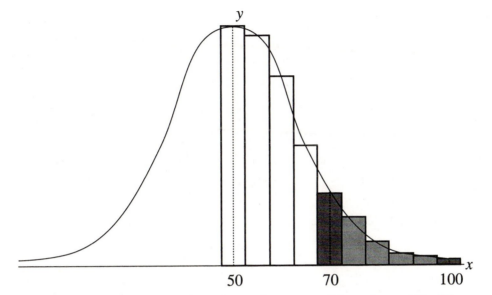

Figure 6–11: Normal Approximation

The area we want includes the rectangles at $x = 70$ and $x = 100$. The rectangle at 70 includes values on the x-axis from 69.5 to 70.5, and the rectangle at 100 includes the values on the x-axis from 99.5 to 100.5. If x is a continuous variable, we want the area under the curve to the right of 69.5 and to the left of 100.5. This is often a confusing point; the important thing to understand is that the following two statements are referencing the same areas:

If x is a discrete variable, we want $70 \leq x \leq 100$.

If x is a continuous variable, we want $69.5 \leq x \leq 100.5$.

Although it looks like we are talking about the same variable we are not. In $70 \leq x \leq 100$, the variable is discrete and it must be a positive integer. In $69.5 \leq x \leq 100.5$, the variable is continuous; there is no longer a restriction that it be an integer. To find the approximate answer to our problem using the normal curve we find

$$P(69.5 \leq x \leq 100.5) \text{ where } \mu = np = 50 \text{ and } \sigma = \sqrt{npq} = 5$$

We now must find the values of the related z scores:

$$z = \frac{69.5 - 50}{5} = 3.9 \qquad z = \frac{100.5 - 50}{5} = 10.1$$

We want the area under the normal curve between 3.9 and 10.1; however, the area to the right of $z = 10.1$ is so small that we can ignore it in this case. We will simply use the area to the right of 3.9; we look up 3.9 in the z table and subtract the value from 0.5 to get our answer:

$$
\begin{aligned}
P(x \geq 70) \text{ for discrete variable} &= P(x \geq 69.5) \text{ for continuous variable} \\
&= 0.5 - 0.49995 \\
&= 0.00005
\end{aligned}
$$

Clearly, if we have either a calculator like the HP-21S or a computer available to get a correct answer, we will not need the normal approximation. But the normal approximation to the binomial distribution provides us with a good opportunity to better understand both distributions. It is because of this opportunity to better understand statistics that we need to spend the time on the approximation even when we have the calculators and computers available.

One more point needs to be addressed regarding the normal approximation. If we had been asked to find the probability of getting more than 70 correct, we could again ignore the area to the right of 100 and write the expression

$$P\left(x_{discrete} > 70\right) = P\left(x_{continuous} \geq 70.5\right)$$

The rectangles for $x > 70$ include rectangles to the right of 70.5 because the first discrete value larger than 70 is 71. It'includes the rectangle from 70.5 to 71.5.

Whenever we have a question as to what value to use in the approximation, it helps to draw the partial probability histogram and label it carefully to see which rectangles we need to use. As we can see from the diagram below, it is clear that we need to use 70.5 for the x value of the continuous variable.

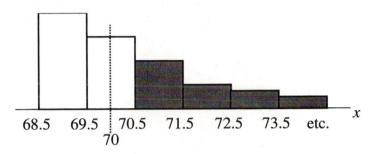

68.5   69.5   70.5   71.5   72.5   73.5   etc. x
                70

Figure 6–12: Continuity Correction Factor

Hot Spot #7 Sample Problem and Solution: Given a 50 question three-part multiple choice test, what is the probability of getting less than 25 correct when guessing? (Include the area to the left of $x = 0$).

Solution: Step 1. $n = 50$ $p = \frac{1}{3}$ and $q = 1 - p = \frac{2}{3}$

Step 2. $np = \frac{50}{3} = 16.\overline{6}$ and $nq = \frac{100}{3} = 33.\overline{3}$
It is OK to use the normal approximation.

Step 3. $\mu = np \approx 16.7$ $\sigma = \sqrt{npq} = \sqrt{\frac{100}{9}} = \frac{10}{3} \approx 3.3$

Step 4.   We want $P(x < 25) = P(x \le 24)$

Step 5.   Draw a partial graph of the histogram and curve to determine the value of the continuous variable x (See Figure 6–15).

Step 6.   We want $P(x \le 24.5)$ and we need to find the related value of $z = \frac{24.5 - 16.7}{3.3} = 2.36$.

Step 7. $P(z \le 2.36)$ $= 0.5 + P(0 \le z \le 2.36)$
$= 0.5 + 0.4949 = 0.9909$

Step 8. $P(x < 25) = P(z \le 2.36) = 0.9909$.

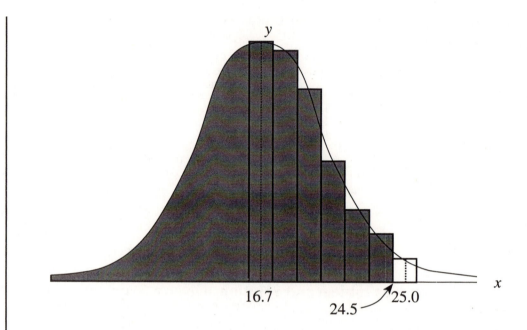

Sample Problem and Answer: Given a binomial random variable with 60 trials and $p = \frac{1}{6}$ and $q = 1 - p = \frac{5}{6}$, find the probability that x is greater than 20. (Include the area to the right of 60).

Answer: $P(x > 20) = P(z \geq 3.64) = 0.0001$

Sample Problem and Answer: Given a 40 question True/False test, find the probability of getting between 20 and 30 (inclusive) questions correct when guessing on each question. Hint: Find the z value for $x = 19.5$ and $x = 30.5$.

Answer: $\mu = 40(0.5) = 20$ $\sigma = \sqrt{40(0.5)(0.5)} = 3.16228$

$z = \frac{19.5 - 20}{3.16228} = -0.16$ $z = \frac{30.5 - 20}{3.16228} = 3.32$

$P(-0.16 \leq z \leq 3.32)$ $= P(-0.16 \leq z \leq 0) + P(0 \leq z \leq 3.32)$

$= 0.0636 + 0.4995$

$= 0.5631$

CHAPTER 6 DISCUSSION QUESTIONS

These questions may be used in your study group or simply as topics for individual reflection. Whichever you do, take time to explain verbally each topic to insure your own understanding. Since these questions are intended as topics for discussion, answers to these questions are not provided. If you find that you are not comfortable with either your answers or that your group has difficulty with the topic, take time to meet with your professor to get help.

1. What is the difference between a discrete and continuous random variable?

2. Why is the probability of a specific value of a continuous random variable equal to zero?

3. What is the standard normal distribution?

4. What is the z score for the mean?

5. What is the relationship of the Empirical Rule to a z score?

6. When finding c so that $P(z \geq c) = 0.05$, how do you decide which side of the bell-shaped curve to place the letter c?

7. Why is it important to graph the curve and locate the values on the curve when finding $P(1.5 \leq z \leq 2.5)$?

8. Given the continuous random variable x with $\mu = 55$ and $\sigma = 10$ why do we need a z score when finding $P(x \geq 57.5)$?

9. What is meant by the phrase "a family of normal distribution curves?"

10. When using the normal approximation to the binomial distribution, what is the correction for continuity?

CHAPTER 6 TEST

1. Use the normal approximation to the binomial distribution to find the probability of more than 35 successes when $n = 40$ and $p = 0.75$

2. Given that x has a normal distribution with $\mu = 85$ and $\sigma = 5.25$, find the probability that x is less than 75.

3. Given that x is a continuous random variable with $\mu = 20.5$, $\sigma = 1.25$, and $n = 49$, find the probability that x is 20.

4. Find the value of z such that the area under the standard normal curve in the left tail is 10%.

5. Find $P(-1.25 \leq z \leq 0.5)$.

6. $P(z \leq -1.96 \text{ or } z \geq 1.96)$.

7. Given the continuous probability distribution defined by $f(x) = \frac{1}{2}x$ for the interval $[0,2]$, find $P(0.25 \leq x \leq 0.75)$.

8. A sample of 100 measurements of temperature are taken from a population whose value of μ is assumed to be $98.2°$. If the sample standard deviation $s = 1.5$, find $P(x = 98.2)$.

9. If a continuous random variable x has a normal distribution with a mean of 35 and standard deviation equal to 5 find $P(x \leq 22.5)$.

10. Use the Empirical Rule to find the probability that test scores will be between 1,150 and 1,300 on a test whose scores are normally distributed with $\mu = 1,200$ and $\sigma = 50$.

CHAPTER 6 TEST Questions and Answers

1. Use the normal approximation to the binomial distribution to find the probability of more than 35 successes when $n = 40$ and $p = 0.75$.

Answer: $\mu = np = (40)(0.75) = 30$,

$\sigma = \sqrt{npq} = \sqrt{(40)(0.75)(0.25)} = \sqrt{7.5} = 2.74$. Use $x = 35.5$,

$z = \frac{35.5 - 30}{2.74} = 2.01$. $P(x > 35) = P(z \geq 2.01) = 0.5 - 0.4778 = 0.0222$.

2. Given that x has a normal distribution with $\mu = 85$ and $\sigma = 5.25$, find the probability that x is less than 75.

Answer: Use $z = \frac{x - \mu}{\sigma}$. $z = \frac{75 - 85}{5.25} = -1.9$.

$P(x < 75) = P(z < -1.9) = 0.5 - 0.4713 = 0.0287$

3. Given that x is a continuous random variable with $\mu = 20.5$, $\sigma = 1.25$, and $n = 49$, find the probability that x is 20.

Answer: $P(x = 20) = 0$

4. Find the value of z such that the area under the standard normal curve in the left tail is 10%.

Answer: This is the same as finding c so that $P(z \leq c) = 10\%$; c must be on the left side of the curve. The z value that gives 40% of the area under the curve between c and 0 is -1.282; So $c = -1.282$.

5. Find $P(-1.25 \leq z \leq 0.5)$.

Answer: $P(-1.25 \leq z \leq 0.5)$ $\qquad = P(-1.25 \leq z \leq 0) + P(0 \leq z \leq 0.5)$

$\qquad\qquad\qquad\qquad\qquad\qquad = 0.3944 + 0.1915$

$\qquad\qquad P(-1.25 \leq z \leq 0.5) \qquad = 0.5859$.

6. $P(z \leq -1.96 \text{ or } z \geq 1.96)$.

Answer: $P(z \leq -1.96 \text{ or } z \geq 1.96) = 0.95$

7. Given the continuous probability distribution defined by $f(x) = \frac{1}{2}x$ for the interval $[0,2]$, find $P(0.25 \leq x \leq 0.75)$.

Answer: $A = (0.5)(2)\left(\frac{2}{2}\right) = 1$. $\qquad A_T = (0.5)(0.75)\left(\frac{0.75}{2}\right) = 0.140625$.

$\qquad\qquad A_t = (0.5)(0.25)\left(\frac{0.25}{2}\right) = 0.015625$.

$\qquad\qquad P(0.25 \leq x \leq 0.75) = 0.140625 - 0.015625 = 0.125$.

8. A sample of 100 measurements of temperature are taken from a population whose value of μ is assumed to be $98.2°$. If the sample standard deviation $s = 1.5$, find $P(x = 98.2)$.

Answer: $P(x = 98.2) = 0$

9. If a continuous random variable x has a normal distribution with a mean of 35 and standard deviation equal to 5 find $P(x \leq 22.5)$.

Answer: $z = \dfrac{22.5 - 35}{5} = -2.5$ and $P(x \leq 22.5) = P(z \leq -2.5) = 0.0062$

10. Use the Empirical Rule to find the probability that test scores will be between 1,150 and 1,300 on a test whose scores are normally distributed with $\mu = 1,200$ and $\sigma = 50$.

Answer: The value of 1,150 is 1 standard deviation to the left of 1,200. The value of 1,300 is 2 standard deviations to the right of 1,200. Using the Empirical Rule 34% of the area is between 1,150 and 1,200. Using the Empirical Rule 47.5% of the area is between 1,200 and 1,300. Therefore, there is $0.34 + 0.475 = 0.815$ or 81.5% of the area between 1,150 and 1,300. Therefore, $P(1,150 \leq x \leq 1,300) = 0.815$.

$\Sigma_\circ$

chapter seven

SAMPLING DISTRIBUTIONS

I know of scarcely anything
so apt to impress the imagination as the wonderful form
of cosmic order expressed by the
"Law of Frequency of Error" (the normal distribution).
— Sir Francis Galton

INTRODUCTION

In chapter 6 we were introduced to a continuous random variable. We then looked at the standard normal distribution and used the z score to find the area under the **standard normal curve**. We were also able to convert a normal distribution involving a continuous random variable x into a standard normal distribution. We then used the z value to find an area under the normal distribution curve.

In this chapter we extend the idea of the probability distribution of a continuous random variable x to the probability distributions of the sample statistics $\bar{x}$ and $\hat{p}$. It is important to note that a sample statistic is a random variable and we refer to the probability distribution of the sample statistic as the **sampling distribution**. Using what is called the **Central Limit Theorem**, we will be able to use the z score to transform the sampling distribution of $\bar{x}$ and $\hat{p}$ into standard normal distributions. The Central Limit Theorem will tell us how to find the **mean and variance of the sampling distribution**. We will then be able to find the probabilities for values of $\bar{x}$ or $\hat{p}$.

As an example, suppose we were interested in the distribution of car weights and want to find the probability that the average weight of 100 randomly selected cars is greater than 2,000 pounds. We need to know a value for the population mean, μ. For the present we will be given the value of the population mean. But later on this value will be obtained from a hypothesis or belief that someone has about the distribution of car weights. Once we have the value of μ, we use the Central Limit Theorem to find the mean and standard deviation of the sampling distribution of $\bar{x}$. We then find the z score for $\bar{x}$ and find the desired probability by finding the area under the right tail of the standard normal curve.

The Central Limit Theorem states that for $n \geq 30$, the sampling distribution of the sample mean is approximately normal irrespective of the shape of the population distribution, and

$$\mu_{\bar{x}} = \mu \text{ and } \sigma_{\bar{x}} = \frac{\sigma}{\sqrt{n}}.$$

The beauty of the Central Limit Theorem is that no matter what the distribution of the continuous variable x, the sampling distribution of $\bar{x}$ will be approximately normal as long as $n \geq 30$.

In a similar manner if np and nq are both greater than 5, the sampling distribution of a sample proportion, $\hat{p}$, is approximately normal, and

$$\mu_{\hat{p}} = p \text{ and } \sigma_{\hat{p}} = \sqrt{\frac{pq}{n}}$$

The steps outlined in the above example on the distribution of car weights become the essence of what we will call hypothesis testing in the chapters that follow. It will be very important that you understand what the Central Limit Theorem states.

CHAPTER 7 HOT SPOTS

1. **The Difference Between** $P(x \geq 65)$ **And** $P(\bar{x} \geq 65)$.

   Starts on **page 7–4**. Problems on **pages 7–8, 7–9, 7–10**.

   Mann 7.5

If you find other HOT SPOTS, write them down and use them as a focus of your discussions in the study group. Or you can use the HOT SPOT as the topic for a help session with your professor.

Hot Spot #1 – The Difference Between $P(x \geq 65)$ And $P(\bar{x} \geq 65)$

One area of difficulty in this chapter is working with the Central Limit Theorem. It is fairly easy to know that the following two statements are part of the CLT:

$$\mu_x = \mu_{\bar{x}} \qquad\qquad \sigma_{\bar{x}} = \frac{\sigma_x}{\sqrt{n}}$$

However, the real problem is to understand the effect these two statements have on a problem we are doing. As an example, we will look at the two questions from our example using the continuous random variable $x =$ the speed of a car where:

$$\mu = 55 \text{ m.p.h. and } \sigma = 5 \text{ m.p.h.}$$

The first question involves finding the probability that a single speed is greater than 65. We write the first question as

$$P(x \geq 65)$$

The second question involves finding the probability that the average speed from a sample of n cars is more than 65. We write the second question as

$$P(\bar{x} \geq 65)$$

In order to understand the difference in these two questions, we need to first look at the distribution of speeds. Figure 7–1 shows the bell-shaped curve. The area shaded represents the 95% of the speeds between 45 m.p.h. and 65 m.p.h. (using the Empirical Rule).

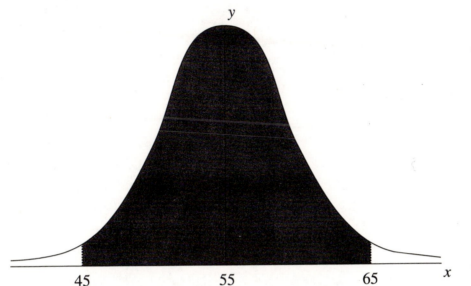

y

45   55   65   *x*

Figure 7–1:
95% Of Speeds
Between
45 M.P.H. and
65 M.P.H.

If we take a sample of speeds from this distribution, it will be unusual to get speeds greater than 65 m.p.h.. In fact, we will only get speeds greater than 65 m.p.h. approximately 2.5% of the time. We can now answer the first question as follows:

$$P(x \geq 65) = 0.025$$

Therefore, 97.5% of the time we would get speeds less than 65 m.p.h.. As a result, the average speed of any sample we take should usually be less than 65 m.p.h.. The difficult part of the question is to quantify the "usually less". This is where we make use of the Central Limit Theorem. The CLT tells us that the standard deviation of the distribution of the $\bar{x}$'s that we get from sampling has a value of 5 m.p.h. divided by the square root of the sample size. If we take a sample of 25 speeds, the standard deviation of the $\bar{x}$'s (sample means) will be $\frac{5}{\sqrt{25}}$.

$$\sigma_{\bar{x}} = 1 \text{ m.p.h..}$$

Sampling Distributions

The bell-shaped curve for the distribution of $\bar{x}$ will be centered at $\mu = 55$, but the standard deviation will now be 1. The curve is shown below: (only the right half of the curve is shown to save space)

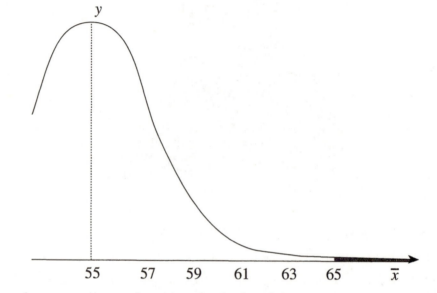

Figure 7–2:
Bell Curve
For Sampling
Distribution
Of Means

As you can see, there is very little of the area under the curve to the right of 65 m.p.h.. We are not able to use the Empirical Rule to find our answer to the second question $P(\bar{x} \geq 65)$, but we can find a z value and use either the z table or the HP-21S calculator to find our answer. The z value is found as follows:

$$z = \frac{65 - 55}{1} = 10$$

Unfortunately our z table only goes up to 5. The probability that z is greater than 5 is given below:

$$P(z \geq 5) = 0.5 - 0.4999997 = 0.0000003$$

Therefore the probability that z is greater than 10 must be smaller than the number we just found. If we use the HP-21S with the key that gives us the area under the normal curve, we find the exact answer of $P(z \geq 10) = 7.62 \times 10^{-24} = 0.0^+$.

A Study Guide For Statistics

We have found that although we might expect to occasionally find one speed selected from the population to be greater than 65, it would be extremely unusual to find a mean of 25 speeds selected from the population to be more than 65. Indeed, this result is the heart of what we do next in the chapter on hypothesis testing. Whenever we get a value that is rare, we consider it to be significant. If we do get such a rare value, we conclude that our original assumption (hypothesis) that μ was 55 m.p.h. is in error and our hypothesis should be rejected.

The two problems we just finished, finding

$$P(x \geq 65) \text{ and } P(\bar{x} \geq 65)$$

make it look like we are working with separate problems, but there is actually a very important common structure to the two problems. Both problems use the z score. We can write the z score in a form that will be very useful as we continue working in the next chapters. We can treat the z score using a template found in spread sheets like LOTUS 1-2-3. We simply fill in the box with the appropriate symbol.

$$z = \frac{[] - \mu_{[]}}{\sigma_{[]}} \text{ where the [] can be } \begin{array}{l} 1.\ x \\ 2.\ \bar{x} \end{array}$$

The same z formula is used for several different z score calculations, with one substitution for each z score.

We are actually using the same formula each time, with the box representing the statistic we are using. Later on we will increase the number of choices for the box to include a difference of sample means, a sample proportion, a difference of sample proportions, and a correlation coefficient. The denominator of the z score is called the standard error of the statistic we are using. When the sample statistic is $\bar{x}$, $\sigma_{\bar{x}}$ is referred to as the standard error of the mean. We often refer to $\sigma_{\bar{x}}$ simply as the standard error.

As we know from the Central Limit Theorem, the original population does not have to be normally distributed to use the theorem for a distribution of sample means. Furthermore, we do not need to know the value of σ if the sample size is considered large, $n \geq 30$. If the sample size is large we can use the value of s, the sample standard deviation, in place of the value of σ. As long as $n \geq 30$ we can write the z score as follows:

$$z = \frac{\bar{x} - \mu}{\frac{s}{\sqrt{n}}}$$

At this point, the use of s for σ may not seem like an important issue, but if $n < 30$ we will have to adjust the formula because s will be an underestimate of σ. We will talk more about this in the chapter that covers small samples using what is called the t distribution.

Hot Spot #1 Sample Problem and Solution: Given a normal distribution of cholesterol readings with $\mu = 200$ and $\sigma = 20$ find the following:

A. $P(x \leq 180)$ B. $P(\bar{x} \leq 180)$ with $n = 100$

Solution:

(A.)   Step 1.   Graph the bell-shaped curve.

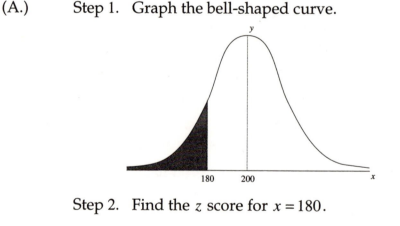

Figure 7-3: Cholesterol Readings

Step 2.   Find the z score for $x = 180$.

$$z = \frac{180 - 200}{20} = -1$$

Step 3. $P(x \leq 180) = P(z \leq -1) = 0.5 - P(-1 \leq z \leq 0)$
$$= 0.5 - 0.3413$$
$$P(x \leq 180) = 0.1587$$

(B.)   Step 1. Graph the bell-shaped curve.

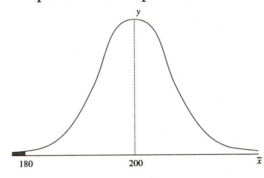

Figure 7–4:
Average
Cholesterol
Readings

Step 2. Find the z score for $\bar{x} = 180$.
$$z = \frac{180 - 200}{\dfrac{20}{\sqrt{100}}} = -10$$

Step 3. $P(\bar{x} \leq 180) = P(z \leq -10) = 0.0^{+}$

Sample Problem and Answer: Given that x has a normal distribution with $\mu = 70$ and $\sigma = 10$ find $P(\bar{x} \geq 75)$ when $n = 100$.

Answer: $P(\bar{x} \geq 75) = P(z \geq 5) = 0.5 - 0.4999997 = 0.0000003$.

Sample Problem and Answer: Given that x has a normal distribution with $\mu = 110$ and $\sigma = 5$ find $P(x \leq 100)$.

Answer: $P(x \leq 100) = P(z \leq -2) = 0.0228$.

CHAPTER 7 DISCUSSION QUESTIONS

These questions may be used in your study group or simply as topics for individual reflection. Whichever you do, take time to explain verbally each topic to insure your own understanding. Since these questions are intended as topics for discussion, answers to these questions are not provided. If you find that you are not comfortable with either your answers or that your group has difficulty with the topic, take time to meet with your professor to get help.

1. What is the sampling error?

2. What does the Central Limit Theorem say about the mean and standard deviation of the random variables $\bar{x}$ and $\hat{p}$?

3. What happens to $\sigma_{\bar{x}}$ and $\sigma_{\hat{p}}$ as n gets larger?

4. How does one control non-sampling errors ?

5. What is an important difference between the population parameter and the random variable $\bar{x}$?

6. Relative to the use of the finite correction factor, when is a sample size considered to be small?

7. What is the finite population correction factor and when should it be used?

8. What is the shape of the sampling distribution of $\bar{x}$ if $n \geq 30$, and and $\hat{p}$ if $np>5$ and $nq>5$?

9. What are the z scores for $\bar{x}$ and $\hat{p}$?

10. What is the difference between a population distribution and a sampling distribution?

CHAPTER 7 TEST

The following information is for problems 1-4: A sample of 100 values of x is randomly selected from a population where x has a normal distribution with $\mu = 85$ and $\sigma = 5.25$.

1. Find the probability that x is less than 75.

2. Find $\mu_{\bar{x}}$ and $\sigma_{\bar{x}}$.

3. Find the probability that $\bar{x}$ is less than 75.

4. What is the sampling error of $\bar{x} = 75$?

5. Given that x is a continuous random variable with $\mu = 20.5$, $\sigma = 1.25$, and a sample of $n = 49$; find the probability that $\bar{x}$ is less than 20.

6. A sample of 100 measurements of temperature are taken from a population whose value of μ is assumed to be $98.6°$. If the sample standard deviation $s = 1.5$, find $P(\bar{x} \geq 99)$.

A sample of 100 students was polled from a population of students where 90% supported extending the time allowed on final exams. Assume that $\frac{n}{N} \leq 0.05$.

7. Find $\mu_{\hat{p}}$ and $\sigma_{\hat{p}}$.

8. Find the z score for $\hat{p} = 85\%$.

9. Find the probability that $\hat{p}$ is less than 85%.

10. What is the sampling error for $\hat{p} = 85\%$?

CHAPTER 7 TEST Questions and Answers

The following information is for problems 1-4: A sample of 100 values of x is randomly selected from a population where x has a normal distribution with $\mu = 85$ and $\sigma = 5.25$.

1. Find the probability that x is less than 75.

Answer: Use $z = \dfrac{x - \mu}{\sigma}$. $z = \dfrac{75 - 85}{5.25} = -1.9$.

$$P(x < 75) = P(z < -1.9) = 0.5 - 0.4713 = 0.0287$$

2. Find $\mu_{\bar{x}}$ and $\sigma_{\bar{x}}$.

Answer: $\mu_{\bar{x}} = \mu = 85$ $\sigma_{\bar{x}} = \dfrac{\sigma}{\sqrt{n}} = \dfrac{5.25}{10} = 0.525$

3. Find the probability that $\bar{x}$ is less than 75.

Answer: Use $z = \dfrac{\bar{x} - \mu}{\dfrac{\sigma}{\sqrt{n}}}$. $z = \dfrac{75 - 85}{\dfrac{5.25}{\sqrt{100}}} = -19.05$.

$$P(\bar{x} < 75) = P(z < -19.05) = 0.0000^{+}$$

4. What is the sampling error of $\bar{x} = 75$?

Answer: The sampling error $= \bar{x} - \mu = 75 - 85 = -10$

5. Given that x is a continuous random variable with $\mu = 20.5$, $\sigma = 1.25$, and a sample of $n = 49$; find the probability that $\bar{x}$ is less than 20.

Answer: Use $z = \dfrac{\bar{x} - \mu}{\dfrac{\sigma}{\sqrt{n}}}$. $z = \dfrac{20 - 20.5}{\dfrac{1.25}{\sqrt{49}}} = \dfrac{-0.5}{0.17857} = -2.8$.

$$P(\bar{x} \leq 20) = P(z \leq -2.8) = 0.5 - 0.4974 = 0.0026$$

6. A sample of 100 measurements of temperature are taken from a population whose value of μ is assumed to be $98.6°$. If the sample standard deviation $s = 1.5$, find $P(\bar{x} \geq 99)$.

Answer: Since $n > 30$, use $z = \dfrac{\bar{x} - \mu}{\dfrac{\sigma}{\sqrt{n}}}$. $z = \dfrac{99 - 98.6}{\dfrac{1.5}{\sqrt{100}}} = 2.67$.

$$P(\bar{x} \geq 99) = P(z \geq 2.67) = 0.5 - 0.4962 = 0.0038$$

A sample of 100 students was polled from a population of students where 90% supported extending the time allowed on final exams. Assume that $\frac{n}{N} \leq 0.05$.

7. Find $\mu_{\hat{p}}$ and $\sigma_{\hat{p}}$.

Answer: $\mu_{\hat{p}} = 0.90$ and $\sigma_{\hat{p}} = \sqrt{\dfrac{pq}{n}} = \sqrt{\dfrac{(.90)(.10)}{100}} = 0.03$

8. Find the z score for $\hat{p} = 85\%$.

Answer: $z = \dfrac{\hat{p} - p}{\sqrt{\dfrac{pq}{n}}} = \dfrac{.85 - .90}{\sqrt{\dfrac{(.9)(.1)}{100}}} = \dfrac{-.05}{.03} = -1.67$

9. Find the probability that $\hat{p}$ is less than 85%.

Answer: $P(\hat{p} < .85) = P(z < -1.67) = 0.0475$

10. What is the sampling error of $\hat{p} = 85\%$?

Answer: The sampling error $= \hat{p} - p = .85 - .90 = -.05$

$\Sigma_\circ$

chapter eight

ESTIMATION OF THE MEAN AND PROPORTION

*When you can measure what you are speaking about
and express it in numbers you know something about it;
but when you cannot measure it,
when you cannot express it in numbers,
your knowledge is of a meager and unsatisfactory kind.*

– Lord Kelvin

INTRODUCTION

In chapter 7, we used the Central Limit Theorem to answer probability questions about a sample statistic. The Central Limit Theorem guaranteed us that when $n \geq 30$, the sampling distribution was approximately normal. We converted values of $\bar{x}$ and $\hat{p}$ into z scores and used the standard normal distribution to answer questions about the normal distribution. Our work in chapter 7 set the stage for what is known as **inferential statistics**.

Inferential statistics involves two key ideas; **estimation** and hypothesis testing. In this chapter we focus on estimation, and in chapter 9 we will focus on hypothesis tests. Briefly stated, estimation involves using sample statistics, such as $\bar{x}$ and $\hat{p}$, to estimate the value of the related population parameters, such as μ and p.

For example, the sample mean $\bar{x}$ is what we call a **point estimate** of the population parameter μ. However, point estimates are not considered very useful because we do not have a way of deciding on the accuracy of our estimate of the population parameter.

There is a second form of estimation that includes both a statement of our confidence in the estimate and the **maximum error of the estimate**. This estimate involves an interval that is constructed around the point estimate. We call this estimate a **confidence interval**. The confidence interval is constructed from either a z or t score using the Central Limit Theorem.

For example with a large sample, we find the **upper and lower limits** of the confidence interval for μ by writing the z score for $\bar{x}$ and solving for the population parameter μ.

$$z = \frac{\bar{x} - \mu}{\frac{s}{\sqrt{n}}} \Rightarrow \mu = \left(\bar{x} \pm z \frac{s}{\sqrt{n}} \right)$$

The value of the z score is determined by the **level of confidence** we want for our interval. For example, if we want to be 90% confident in our estimate of μ, we use a z score that gives

$$\frac{100\% - 90\%}{2} = 5\%$$

of the area under the standard normal curve to the right of z. In the case of the 90% confidence interval for μ, we use $z = 1.65$. So we would say that we were 90% confident that μ would be in the interval

$$\left(\bar{x} - 1.65 \frac{s}{\sqrt{n}}, \bar{x} + 1.65 \frac{s}{\sqrt{n}} \right)$$

We finish the chapter by focusing on the size of the sample we need to limit the error of our estimate of the population parameter.

CHAPTER 8 HOT SPOTS

1. **The Difference Between A Point And An Interval Estimate**

   Starts on **page 8–4.** Problems on **pages 8–8, 8–9.** Mann 8.1, 8.2

2. **Degrees Of Freedom.**

   Starts on **page 8–10.** Mann 8.4.1

If you find other HOT SPOTS, write them down and use them as a focus of your discussions in the study group. Or you can use the HOT SPOT as the topic of a help session with your professor.

Hot Spot #1 – The Difference Between A Point And An Interval Estimate

Earlier in the text we used the sample mean, $\bar{x}$, to estimate the population mean, μ. The sample mean is called a point estimate of the population parameter. We need to take a sample from the population to obtain the data we use to calculate the sample mean. If the size of the sample is large, we feel good about our estimate. On the other hand if the sample size is small, we may feel uneasy about our point estimate of the parameter. In either situation, we do not have any way of finding out how accurate we are in our estimate.

In contrast, the confidence interval gives us a way of estimating a population parameter that includes a measure of how confident we are and the maximum error of our estimate. Constructing a confidence interval involves the same ideas we used when we worked with the Empirical Rule with a known value of μ to find an interval that contained $\bar{x}$ 95% of the time. We simply reverse the process in finding a confidence interval. We start with a known value of $\bar{x}$, and find an interval that will contain the true value of μ 95% of the time.

We will first look at using the Empirical Rule with the results of the Central Limit Theorem to find an interval that contains the sample mean, $\bar{x}$, 95% of the time. In this case we know the value of μ. As an example let x be a continuous random variable that represents SAT scores for a normal population with

$$\mu = 1,200 \text{ and } \sigma = 100 \text{ and } n = 100.$$

The Empirical Rule tells us that 95% of the sample means should be in the interval

$$\mu \pm \frac{2\sigma}{\sqrt{n}}.$$

The interval includes sample means from $1,200 - \dfrac{2(100)}{\sqrt{100}}$ to $1,200 + \dfrac{2(100)}{\sqrt{100}}$

The interval is $(1{,}180\,,1{,}220)$ and is illustrated in the following graph:

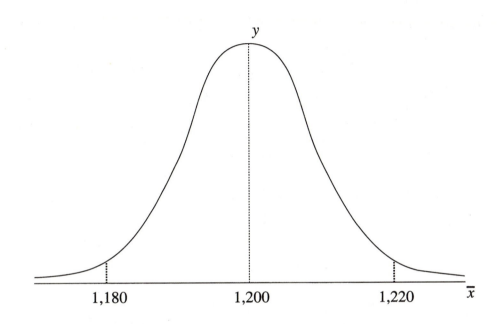

Figure 8–1:
Interval from
1,180 to 1,220

In contrast, when we construct a confidence interval, we do not know the value of μ. So we find $\bar{x}$ and use the idea of a z score to create an interval that contains μ. We start with the z score:

$$z = \frac{\bar{x} - \mu}{\dfrac{\sigma}{\sqrt{n}}}$$

We want what we will call the 95% confidence interval for estimating the population mean, μ. We will be 95% confident that the interval we construct from $\bar{x}$ will contain the true value of μ. In this sense we will accept being wrong 5% of the time. The z score for the 95% confidence interval is found by getting a value for $z\left(\frac{0.05}{2}\right)$. When $z = 1.96$, we have 2.5% of the area under the normal curve in the right tail. So we will use $z \pm 1.96$ to get our 95% confidence interval.

We now substitute for z and rewrite the z score. Solving for μ we get the following:

$$\mu = \bar{x} \pm 1.96 \frac{\sigma}{\sqrt{n}}$$

Generally speaking, we do not know σ. So, we use the sample standard deviation s as an estimate for σ. Since we are using the Central Limit Theorem to find $\sigma_{\bar{x}}$, it is important that the sample size be large, $n \geq 30$.

Returning to our example on SAT test scores, we will consider the case where we have no idea of the value of μ. As an illustration we will assume that all we have is a sample with

$$\bar{x} = 1{,}210 \text{ and } s = 100 \text{ and } n = 100.$$

We want to find the 95% confidence interval for μ. So we substitute the values of $\bar{x}$, s, and n into the formula as follows:

formula

- $\mu = \bar{x} \pm 1.96 \dfrac{\sigma}{\sqrt{n}}$

substituting

- $\mu = 1{,}210 \pm 1.96 \dfrac{100}{\sqrt{100}}$

simplifying

- $\mu = 1{,}210 \pm 19.6$

interval

- $1{,}190.4 \leq \mu \leq 1{,}229.6$

We then say that we are 95% confident that the interval from 1,190.4 to 1,229.6 contains the population mean μ.

We usually do not know if the interval actually contains the true value of μ. We can only say we are 95% confident that the interval contains μ. However, to see how the confidence interval works, we assumed we knew that $\mu = 1{,}200$. In the above example, the confidence interval we constructed did contain $\mu = 1{,}200$.

Now, let us see what it means to be 95% confident that the interval contains μ. We need to return to what we did with the Empirical Rule, where we used a value of 2 rather than the more accurate z score of 1.96. We found that if $\mu = 1{,}200$ then 95% of the sample means would be in the interval from 1,180 to 1,220. If we use any

$\bar{x}$ in this interval to construct the confidence interval for μ, the confidence interval will contain μ.

The 95% confidence interval for μ we constructed from $\bar{x} = 1,210$ contained the value of 1,200. We need to note that $\bar{x}$ was in the interval from 1,180 to 1,220 that we obtained from the Empirical Rule. In fact, if we construct the confidence interval using any $\bar{x}$ from the interval $(1,180 , 1,220)$, the confidence interval will contain the value 1,200.

Only 5% of the $\bar{x}$'s are outside of the interval from 1,180 to 1,220. If we use a $\bar{x}$ outside the interval $(1,180 , 1,220)$, the confidence interval we construct will not contain 1,200. To show this we will use $\bar{x} = 1,175$ and $s = 100$ and $n = 100$.

- $$\mu = \bar{x} \pm 1.96 \frac{\sigma}{\sqrt{n}}$$ *formula*

- $$\mu = 1,175 \pm 1.96 \frac{100}{\sqrt{100}}$$ *substituting*

- $$\mu = 1,175 \pm 19.6$$ *simplifying*

- $$1,155.4 \leq \mu \leq 1,194.6$$ *interval*

This time the 95% confidence interval goes from 1,155.4 to 1,194.6 and the confidence interval does **not** contain $\mu = 1,200$.

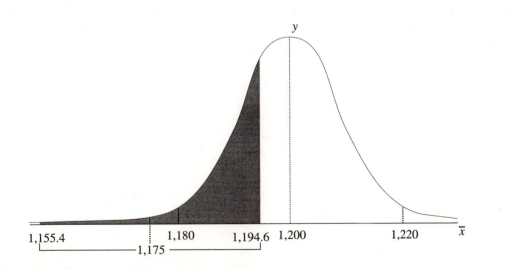

Figure 8–2: Illustration Of Choosing An $\bar{x}$ Outside The 95% Confidence Interval

The following table helps summarize the results of the examples:

Table 8–1:
Summary Of
Analysis Of
Confidence
Interval

Summary
1. $\bar{x}$ is in the interval $(1{,}180\,,1{,}220)$ 95% of the time.
2. We use an $\bar{x}$ in the interval to construct the confidence interval.
3. The confidence interval contains $\mu{=}1200$.
1. $\bar{x}$ is **not** in the interval $(1{,}180\,,1{,}220)$ 5% of the time.
2. We use an $\bar{x}$ outside the interval to construct the confidence interval.
3. The confidence interval does **not** contain $\mu{=}1200$.

Another way of saying what is in the table is that if:

1. We take 100 different random samples, with $n \geq 30$;
2. Find $\bar{x}$ and s for each sample;
3. Construct the confidence interval for each $\bar{x}$, then 95 of the 100 confidence intervals will contain $\mu = 1{,}200$.

Hot Spot #1 Sample Problem and Solution: Given the distribution of SAT scores discussed above with $\bar{x} = 1{,}210$, $s = 100$, and $n = 100$ show that the 95% confidence interval contains $\mu = 1{,}200$.

Solution: Step 1. Substitute values of $\bar{x}$, s, and n into $\mu = \bar{x} \pm 1.96 \dfrac{\sigma}{\sqrt{n}}$.

$$\mu = 1{,}210 \pm 1.96 \frac{100}{\sqrt{100}}$$

Step 2.   Determine the upper and lower values of the interval
$\mu = 1,210 \pm 19.6$.

$1,190.4 \le \mu \le 1,229.6$

Step 3.   The interval from 1,190.4 to 1,229.6 contains $\mu = 1,200$.

Sample Problem and Answer: Given the distribution of SAT scores discussed above with $\bar{x} = 1,250$, $s = 100$, and $n = 100$ show the 95% confidence interval does **not** contain $\mu = 1,200$.

Answer:

• $\mu = \bar{x} \pm 1.96 \dfrac{\sigma}{\sqrt{n}}$ *formula*

• $\mu = 1,250 \pm 1.96 \dfrac{100}{\sqrt{100}}$ *substituting*

• $\mu = 1,250 \pm 19.6$ *simplifying*

• $1,230.4 \le \mu \le 1,269.6$ *interval*

The confidence interval does **not** contain $\mu = 1,200$.

Sample Problem and Answer: Given the following five values of $\bar{x}$ for the SAT scores discussed above, a 95% confidence interval is to be constructed for each $\bar{x}$. Which one of the values of $\bar{x}$ will result in the confidence interval containing $\mu = 1,200$?

| 1,195 | 1,235 | 1,218 | 1,100 | 1,215 |

Answer: All the values, except 1,100 and 1,235, will result in a 95% confidence interval that contains $\mu = 1,200$.

Hot Spot #2 – Degrees Of Freedom

The t distribution requires that we find what are called degrees of freedom before we can find critical values for hypothesis testing. The letters df are used to represent the degrees of freedom. In both the t and chi-square distributions the degrees of freedom value is equal to one less than the sample size.

$$df = n - 1$$

The formula makes it simple to find the degrees of freedom, but it does not provide us any insight into what the phrase means. The idea behind degrees of freedom is illustrated by the following example:

- We want 3 numbers that have a sum of 10.

- We are free to choose any two numbers we want. For example we could choose 2 and 3.

- The third number must be a number that provides a sum of 10. In our example, the third number must be a 5. In this sense we do not have any freedom relative to the third number.

In our example we only had two degrees of freedom,

$$df = 3 - 1 = 2$$

If we had started with five numbers that had a sum of 10, we would have been free to choose any four numbers. However, the fifth number would not be a free number. Once we select four numbers, the fifth number is determined by the sum. The degrees of freedom would be $5 - 1$.

In the same way, if we started with n numbers that have a sum of 10, we are only free to choose $n - 1$ numbers. The last number is again determined by the sum; so, degrees of freedom equals $n - 1$.

The important point about the above examples is the effect that the sum 10 had on the numbers we selected. We used a sum of 10 to illustrate what happens, but any sum could have been used. For that matter we could have used the mean rather than a sum. The restriction that the numbers have a particular mean has the same effect of reducing our degrees of freedom to $n-1$.

In fact we have a similar loss of freedom for any set of numbers when the numbers are used in a calculation that must have a given value. It does not matter what calculation is done. The effect is the same. The sum was used in the above example because it is an easy calculation to do. The mean was used as a second example of a calculation that reduced the degrees of freedom because it involves a sum. However, we would have the same loss in degrees of freedom if we used the restriction that the numbers we select have a particular standard deviation.

One final point needs to be made about degrees of freedom that will be used in later chapters. The point involves what happens when we select numbers if more than one calculation is involved as a restriction. For example, if the numbers we select must be used in two calculations, we lose two degrees of freedom and write $df = n-2$. We can illustrate the idea using a familiar example from beginning algebra to show how we would lose two degrees of freedom. The problem is usually done with a variable and an equation that is then solved. In this case the problem is easy enough to solve that we will just give the solution by using observation.

- We want two numbers that have a sum of 10 and a product of 24.

- There are only two numbers that meet the requirement. The numbers are 6 and 4. We do not have any freedom. We started with two numbers, but lost two degrees of freedom because the numbers had to have a particular sum and product. Therefore, $df = 2-2 = 0$.

The above idea can be generalized in the following way:

> If the numbers we select must be used in k calculations that
> restrict the numbers we select, we lose k degrees of freedom and
> write $df = n - k$.

For example, if we had a large sample that we divided into k groups
and required that the mean of each group be a specific value, we then
would have $n - k$ degrees of freedom.

CHAPTER 8 DISCUSSION QUESTIONS

These questions may be used in your study group or simply as topics for individual reflection. Whichever you do, take time to explain verbally each topic to insure your own understanding. Since these questions are intended as topics for discussion, answers to these questions are not provided. If you find that you are not comfortable with either your answers or that your group has difficulty with the topic, take time to meet with your professor to get help.

1. What is the difference between a point and an interval estimate?

2. What is the error of the estimate?

3. How do you reduce the error of the estimate for μ?

4. What is meant by degrees of freedom?

5. How do you decide on the value of the z or t score to use with the confidence interval?

6. How does a t value compare to a z value?

7. How do we narrow the width of the confidence interval?

8. How is the Central Limit Theorem used in constructing confidence intervals?

9. When do we use s to find the upper and lower limits of the confidence interval?

10. Why do we use $p = q = 0.5$ when we find the sample size for a given error when estimating the population proportion?

CHAPTER 8 TEST

1. Given $\bar{x} = 1{,}225$, $s = 35$ and $n = 25$, find the point estimate of μ.

2. Given $\bar{x} = 1{,}225$, $s = 35$ and $n = 25$, find the 95% confidence interval for μ.

3. What is the t value for a 95% confidence interval with n=20?

4. If you want a smaller confidence interval for μ, what two choices do you have?

5. Given a sample of 50 scores with $\bar{x} = 1{,}237$ and $s = 125$, find the the value of E, the maximum error of the estimate, for the 95% confidence interval for the population mean.

6. What sample size is necessary to estimate the population proportion within 0.01 with 95% confidence?

7. A random survey of 100 managers found that 35 of them favored a national advertising campaign, find a 95% confidence interval for the population proportion, p.

8. What values of p and q should be used to find the sample size for a given value of E?

9. What is the value of α for a 95% confidence interval?

10. Which of the following confidence levels will give the narrowest interval: 90% or 95%?

CHAPTER 8 TEST Questions and Answers

1. Given $\bar{x} = 1,225$, $s = 35$ and $n = 25$, find the point estimate of μ.

Answer: $\bar{x} = 1,225$ is the point estimate of μ.

2. Given $\bar{x} = 1,225$, $s = 35$ and $n = 25$, find the 95% confidence interval for μ.

Answer:
$$t(0.975) = -2.064 \qquad t(0.025) = +2.064$$
$$1,225 - \frac{(2.064)(35)}{\sqrt{25}} < \mu < 1,225 + \frac{(2.064)(35)}{\sqrt{25}}$$
$$1,210.552 < \mu < 1,239.448$$

3. What is the t value for a 95% confidence interval with $n = 20$?

Answer: $t = 2.093$

4. If you want a smaller confidence interval for μ, what two choices do you have?

Answer: Either increase the sample size or increase the value of α.

5. Given a sample of 50 scores gave $\bar{x} = 1237$ and $s = 125$, find the value of E, the maximum error of the estimate, for the 95% confidence interval for the population proportion, p.

Answer: $E = 1.96 \left(\frac{125}{\sqrt{50}} \right) = 34.65$.

6. What sample size is necessary to estimate the population proportion within 0.01 with 95% confidence?

Answer: $n = \left[\dfrac{1.96}{0.01}\right]^2 \cdot \dfrac{1}{4} = 9,604$.

7. A random survey of 100 managers found that 35 of them favored a national advertising campaign. Find a 95% confidence interval for the population proportion, p.

Answer: $\hat{p} = 0.35$

$$\hat{p} - 1.96\sqrt{\dfrac{\hat{p}\hat{q}}{n}} \le P \le \hat{p} + 1.96\sqrt{\dfrac{\hat{p}\hat{q}}{n}}$$

$$0.35 - 1.96\sqrt{\dfrac{(0.35)(0.65)}{100}} \le P \le 0.35 + 1.96\sqrt{\dfrac{(0.35)(0.65)}{100}}$$

$$0.2565 \le P \le 0.4435.$$

8. What values of p and q should be used to find the sample size for a given value of E?

Answer: $p = q = 0.5$

9. What is the value of α for a 95% confidence interval?

Answer: $\alpha = \dfrac{100\% - 95\%}{2} = 0.025$

10. Which of the following confidence levels will give the narrowest interval: 90% or 95%?

Answer: 95%

$\Sigma_\circ$

chapter nine

HYPOTHESIS TESTS ABOUT THE MEAN AND PROPORTION

It is a capital mistake to theorize
before one has data.

– Arthur Conan Doyle

INTRODUCTION

In the last three chapters we worked with continuous probability distributions. The z score was first presented in chapter 6. We used the standard normal variable z and the related t distribution for small samples to answer probability questions related to the distributions of x, $\bar{x}$ and $\hat{p}$. In chapter 8 we used the z and t scores to extend our work to confidence intervals and estimating the sample size to obtain a degree of accuracy in the interval estimate of the population mean and proportion.

The **confidence interval** was an extension of the work we did in chapter 3, where we found what are called **point estimates** of the population parameter. For example $\bar{x}$ was a point estimate of the population mean. We used both $\bar{x}$ and the z score to find an interval that contained μ. More importantly, we were able to state a level of confidence that the interval we created contained μ a certain percent of the time.

This chapter extends our work to what is described as **hypothesis testing.** The idea is simply an extension of what we did in the last two chapters with the z and t score. For example, we used the z score to find the probability that the sample mean was greater than some value. In each problem we were given the value of μ and σ to use in our calculation of z. We then used the z table to find the probability.

In this chapter we are given μ or p as a hypothesized value someone believes to be true. We start by first finding the value of the z score or value of t related to the sample statistic $\bar{x}$ or $\hat{p}$ and the hypothesized value of the population parameter. We then find the probability that the **test statistic** is larger than or less than a **critical value** found in the z table or given by a calculator.

If the probability we find is less than 5% (which is called the level of significance, α), we decide to reject the **null hypothesis.** The value of α states the percent of the time that we are wrong if we reject a null hypothesis that is true. There are several new terms defined that relate to hypothesis testing. It is important to take the time to learn what the terms mean.

An important key to this chapter, and the chapters that follow, is the general structure of hypothesis testing. This process will be used in all of the remaining chapters and involves five parts:

1. State the **null and alternate hypotheses,** H_o and H_a.
2. Select the **distribution** to use.
3. Determine the **rejection and nonrejection regions** using a table.
4. Calculate the value of the **test statistic.**
5. Make a decision to either **reject H_o** or **fail to reject H_o.**

A Study Guide For Statistics

CHAPTER 9 HOT SPOTS

If you find other HOT SPOTS, write them down and use them as a focus of your discussions in the study group. Or you can use the HOT SPOTS as the topic for a help session with your professor.

Hot Spot #1 – Deciding On The Null Hypothesis, H_0.

It is fairly easy to decide on the null hypothesis when doing our own research. All we do is let the alternate hypothesis equal what we want to prove. We often refer to the alternate hypothesis as the research hypothesis. The null hypothesis then becomes the complement of the alternate hypothesis or research hypothesis. For example, if we want to prove that the population mean of highway speeds is greater than 55, we write

$$H_a: \quad \mu > 55.$$

The null hypothesis then becomes the complement of > 55.

$$H_0: \quad \mu \leq 55.$$

We usually write the null hypothesis using the = sign as follows: (**Hot Spot #4** discusses why we use the equal sign).

$$H_0: \quad \mu = 55.$$

Technically, a null hypothesis can be written with any one of the following three symbols:

$$\leq \quad \text{or} \quad = \quad \text{or} \quad \geq$$

Rather than making it difficult to determine which form to use in the null hypothesis, as a rule of thumb, we just remember that **the null hypothesis has the equal sign in it**.

The alternate hypothesis (called the research hypothesis) then has one of the following symbols in it:

$$< \quad \text{or} \quad \neq \quad \text{or} \quad >$$

(**Hot Spot #3** discusses how to decide on the choice of symbols for the alternate hypothesis). The choice of the alternate hypothesis really is just a matter of deciding whether the critical region is in one or two tails of the standard normal curve.

Remember: The null hypothesis has the equal sign in it.

Hot Spot #2– Deciding On The Test Statistic

✳
*Mann
Sections 9.2,
9.4, And 9.5*

In this chapter, we use the test statistic z in our test of hypothesis. The sample statistic we use with the z test can be in one of three forms. We can use the sample mean ($\bar{x}$), the sample proportion ($\hat{p}$), or the number of successes (x). The choice of the test statistic is made as we read the problem. We are looking for two words, mean and proportion. When a problem involves the proportion we often have to interpret the meaning from the way the problem is written. One of the keys is to look for a percent that is given; the value given will usually be the value of the population proportion used in H_0.

When we read that we are trying to show something is significant about the population mean, we use

$$z = \frac{\bar{x} - \mu}{\frac{\sigma}{\sqrt{n}}}$$

If we read that we are trying to show something significant about the population proportion, we use

$$z = \frac{x - np}{\sqrt{npq}} \quad \text{or} \quad z = \frac{\hat{p} - p}{\sqrt{\frac{pq}{n}}}$$

The second of these two formulas for z has the advantage of fitting the structure we used earlier:

$$z = \frac{[] - \mu_{[]}}{\sigma_{[]}} \text{ where the } [] \text{ can be } \begin{array}{l} 1.\ x \\ 2.\ \bar{x} \\ 3.\ \hat{p} \end{array}$$

*Expanded
Template For z
Test Statistic*

We use the z test statistic to make inferences about the population parameter. As we go into the following chapters we will continue to use the z value to test hypotheses about the population parameter.

Hot Spot #3 – Deciding On A One-Tailed Versus A Two-Tailed Test

The research hypothesis is what we want to prove. We write the research hypothesis as what we call the alternate hypothesis, H_a. We use one of the following three symbols of inequality in H_a:

$$< \quad \text{or} \quad \neq \quad \text{or} \quad >$$

We look for key words in the problems we do to decide on the choice of the inequality symbol to use with H_a. The following table shows many of the words that are used with each of the inequality symbols. In each situation the statement in the problem would involve showing that the population parameter was significantly _____, where the blank would be one of the words taken from the table below:

Table 9–1:
Words And
Symbols Used In
The Alternative
Hypothesis

Table of Words and Symbols Used in H_a

$<$	$\neq$	$>$
Less Than	Different	More Than
Decreased	Changed	Increased
Reduced		Greater
Lowered		Higher
Worse		Better

One of the important points to remember is that the direction of the inequality symbol tells us the location of the critical region. The inequality symbol is like the directional signal in a car. When the arrow points left <, we have the critical region on the left side of the curve. When the arrow points right >, we have the critical region on the right side of the curve. When the symbol is not equal ≠, it is like having the emergency flasher system on where both tail lights blink. We then have the critical region in both tails of the curve.

Hot Spot #4 – Why We Use The = In A One-Tailed Test

✳
*Mann
Section 9.1.4*

It is easiest to look at this question through an example using highway speeds. In a hypothesis test of $H_{0:}\ \mu \leq 55$ and $H_{a:}\ \mu > 55$ with $\sigma = 5$, a sample of $n = 25$ gave a sample mean of 57. We are interested in the probability that $\bar{x}$ is greater than 57 given that $\mu \geq 55$. We can start by looking at the normal curve with $\mu = 55$. Using the z formula for $\bar{x} = 57$, we get $z = 2$. The area to the right of 57 is $P(z > 2) = 0.0228$, and represents 2.28% of the area under the curve. The probability equal to the area under the tail of the curve is called the significance level of the hypothesis test. We use the symbol α for the size of the significance level.

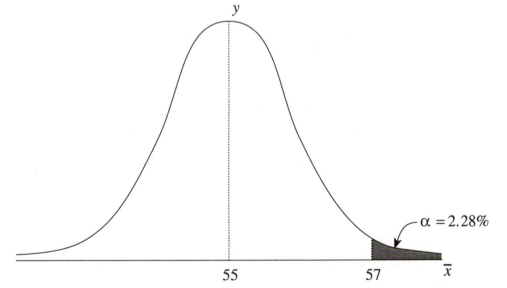

*Figure 9–1:
Significance
Level Of The
Hypothesis Test*

Next, we will look at the area under the normal curve for a value of $\mu < 55$, that is $H_{0:}\ \mu = 53$. Using the z formula for $\bar{x} = 57$, we get $z = 4$. Using the HP-21S, we find that the area to the right of 57 is $P(z > 4) = 0.0000317$ and represents 0.00317% of the area under the curve. This time we say that $\alpha = 0.0032\%$. The table in Table 7–2 below shows the value of α for different values of μ using the HP-21S.

Values of α for Different Population Mean Values	
μ	α
55	2.28%
54	0.13%
53	0.0032%
52	0.000029%
51	0.00000010%

Table 9–2:
Significance
Levels For
Different
Population Mean
Values

When we compare values in the table, we see that for values of $\mu < 55$, the area under the tail of the curve is much less than the area for $\mu = 55$. In this sense, the value of μ that is given by the equal sign in the null hypothesis, H_0, gives us the worst case scenario for the α value; that is, the largest value of α. Thus, it is common practice to simply write the null hypothesis with the equal sign. In the above example we would write

$$H_0: \mu = 55 \text{ and } H_a: \mu > 55.$$

✳
Mann
Section 9.1.3

Hot Spot #5 – Choosing A Value Of α

It is important that we recall that the value of alpha (α) is the probability of rejecting a true null hypothesis. When we reject H_0, we report our findings to others in the form of a research paper or news release. If the null hypothesis is true and we make the mistake of rejecting H_0 and reporting our results to the world, we will be very embarrassed to say the least. Most of us are very worried about making a fool of ourselves in public. So we try to keep the value of α small. We usually choose a value of 1% or 5% for α. There are times that we may select a value of 10% when we are in the early stage of a research program. It is often difficult to find significant results, and we may be willing to accept a large value of α.

There are other considerations that may affect the value of α. For example, if we plan to publish a research report in a journal, it may be a common and accepted practice that all research be conducted using a value of 5%. In other situations, we may be doing a research project and a committee or advisor may decide on the level of α that should be used.

Another point is involved when we choose the value of α. We will illustrate the point using a right-tailed hypothesis test. The smaller the value of α we choose, the larger the critical value. For example, in a one-tailed test using the z test statistic, the critical value is 1.645 for $\alpha = 5\%$, but it is 2.326 when $\alpha = 1\%$. In order to reject H_0 at 1%, we need a larger value of z. Since

$$z = \frac{\bar{x} - \mu}{\sigma_{\bar{x}}}$$

we will need a larger difference between $\bar{x}$ and μ. In research it is usually very difficult to find large differences between the sample mean and its hypothesized value. Thus, it makes more sense to use the smaller critical value of 1.645 to make it easier to reject H_0.

It helps to keep in mind the inverse relationship between α and the critical value:

α gets smaller as the critical value gets larger.

In summary, if we must choose the value of α on our own, we use 5%.

Hot Spot # 6 – The Relationship Between α And The Reported P Value.

There is a common practice these days to actually report the significance level at the value given by the test statistic. The following example illustrates the practice:

In a hypothesis test of $H_0: \mu = 55$ and $H_a: \mu > 55$ with $\sigma = 5$, a sample of $n = 25$ gave a sample mean of 57. Test for a significant increase at $\alpha = 5\%$.

The value of the test statistic z is

$$z = \frac{57 - 55}{\frac{5}{\sqrt{25}}} = 2$$

We then find the probability that z is greater than 2:

$$P(z > 2) = 0.0228$$

We say that the reported P value of α is 2.28%.

We now have several choices that we can make:

1. Since the critical value is 1.645, we can report our research findings significant at the predetermined value of $\alpha = 5\%$.

2. We can simply report our research finding at the reported P value of 2.28%.

3. We can combine the two approaches by arguing that the P value of 2.28% is less than $\alpha = 5\%$; hence, we reject Ho.

There are two disadvantages to the first choice. First, we are not able to pass on the additional information that the reported P value provides, namely how significant our result is. Secondly, there are times when a reported P value would allow us to report research findings that would otherwise go unreported. For example, if the sample mean in the above hypothesis test was equal to 56.5, the value of the test statistic would have been equal to 1.5; we would fail to reject at $\alpha = 5\%$. However, if we decide to use the reported P value, we find that the probability that z is greater than 1.5 is 0.0668. We could report a significant increase at 6.68% and leave it to the reader to determine if he or she wishes to accept the risk of the type I error.

In the second choice, we seem to have the advantage of not having to worry about a predetermined significance level. However, there is another consideration that we need to address. The value of α (the probability of rejecting a true null hypothesis) is related to the value of β (the probability of failing to reject a false null hypothesis). The two values are inversely related.

$$\alpha \text{ goes down in value as } \beta \text{ goes up in value.}$$

Generally speaking, we do not worry too much about the value of β when we fail to reject the null hypothesis.

The type I error associated with α is considered the more serious error. However, β is related to the power of the test. The power of the hypothesis test is the probability of rejecting a false null hypothesis.

$$\text{Power} = 1 - \beta$$

As β gets larger, the power gets smaller. It should be clear that very small values of α ultimately affect the power. As α gets smaller, β gets larger, and power gets smaller.

It is not appropriate to provide quantitative comparisons of the three values at this point in the course. But it is important to understand that one should not allow α to be reported at very low values without considering the possible effect on the power of the test.

In the third choice we are able to combine the best of what is involved in the first and second choices. When we use the third choice, it is common practice to first state the significance level of the hypothesis test at $\alpha = 5\%$. We then find the reported P value and compare it to α. If $P < \alpha$ then we reject H_0. The reported P value then has the advantage of showing how significant the result is.

One of the more subtle advantages of using the P value with a stated level of significance is that we do not need to find critical values. If we are using $\alpha = 5\%$, then any value of $P < 5\%$ will allow us to reject H_0. Note that in the traditional approach to hypothesis testing, we reject when $|z|$ is large, but using the P value we reject H_0 when P is small.

Mann
Sections
9.3 And 9.4

Hot Spot #7– Reporting P Values For The t Distribution

Reporting P values for a t distribution can be a problem when we do not use a calculator like the HP-21S. If we use the calculator, we can easily find the reported P value. All we have to do is enter the calculated value of t and the value for the degrees of freedom and press one key to give us the reported P value. However, P values can be a problem without the calculator.

The reason for the problem is that the table we use for critical values of t lists all the values of t for each degree of freedom under 30 for each value of α. The following table shows a typical row of values in a t table for $df = 9$:

df	$t(0.005)$	$t(0.01)$	$t(0.025)$	$t(0.05)$	$t(0.10)$
9	3.250	2.821	2.262	1.833	1.383

The problem with reporting a P value for a calculated t test statistic is shown in the following example where a hypothesis test was used to find a significant reduction in cholesterol levels for a sample of

10 patients who reduced the amount of fat in their diets. The information used in the hypothesis test was as follows:

$$H_0: \mu = 200 \quad \text{and} \quad H_a: \mu < 200$$

$$\bar{x} = 192 \quad s = 11.35 \quad n = 10$$

$$t = \frac{192 - 200}{\frac{11.35}{\sqrt{10}}} = -2.229$$

Since the alternate hypothesis points to the left side of the curve, we need to think of the values in the above table as representing negative critical values on the left side of the curve. We then compare the calculated (observed) value of 2.229 to the entries in the table. Our observed value is between the critical values for α at $2\frac{1}{2}\%$ and 5%.

$$-2.262 < -2.229 < -1.833$$

We report the significance level as a value of less than 5% but more than $2\frac{1}{2}\%$.

$$2\frac{1}{2}\% < \alpha < 5\%$$

Using the HP-21S, we find that the actual P value is 0.0264.

Hot Spot #7 Sample Problem and Solution: A hypothesis test was done using the information that follows. Find the reported P value.

$$H_0: \mu = 1,200 \quad \text{and} \quad H_a: \mu > 1,200$$

$$\bar{x} = 1,214.75 \quad s = 23.14 \quad n = 20$$

$$t = \frac{1,214.75 - 1,200}{\frac{23.14}{\sqrt{20}}} = 2.851$$

Solution: Step 1. We calculate $df = 20 - 1 = 19$.

Step 2. We refer to the t table at 19 degrees of freedom and find the calculated value of t lies between the critical values of 2.539 at 1% and 2.861 at 0.5%.

Step 3. We report the significance level as less than 1% and greater than 0.5% and write $0.5\% < \alpha < 1\%$.

Sample Problem and Answer: A hypothesis test of $H_0: \mu = 55$ and $H_a: \mu \neq 55$ with $n = 10$ resulted in a value of $t = 1.842$. What is the reported P value?

Answer: Using $df = 9$, we find the calculated t value between 1.833 at 5% and 2.262 at $2\frac{1}{2}\%$. Since the hypothesis test was two-tailed we say that $\frac{\alpha}{2}$ is between $2\frac{1}{2}\%$ and 5%. So the hypothesis test was significant at a value of α between 5% and 10%.

Sample Problem and Answer: A hypothesis test of $H_0: \mu = 220$ and $H_a: \mu > 220$ with $n = 30$ resulted in a value of $t = 2.284$. What is the reported P value?

Answer: Using $df = 29$, we find the calculated t value between 2.045 at $2\frac{1}{2}\%$ and 2.462 at 1%, we say that the hypothesis test was significant at a value of α between 1% and $2\frac{1}{2}\%$.

Hot Spot #8 – The Difference Between Reject And Fail To Reject

✳
Mann
Section 9.1.2, 9.2

It is fairly easy to figure out what is means to reject the null hypothesis. Failing to reject is just the flip side of the coin. The statement "Fail To Reject" is **not**, however, the same as accepting the null hypothesis. When we reject H_0, we do so at the significance level called alpha, α. The significance level states the percentage of the time we are making a type I error by rejecting H_0, when it is in fact true.

When we make the statement that we "Fail To Reject H_0, we are not obligated to give an error analysis. We have not in effect made any decision about the null hypothesis. If we, however, decided to accept H_0, we would need to indicate the percentage of the time we are making an error. The error in accepting a false H_0 is called a type II error, and the probability of making the error is called beta, β. Even though we agree on a significance level (α) at the beginning of a hypothesis test, we seldom know the value of β. Therefore, when the test statistic does not fall in the critical region, the best we can do is say that we fail to reject H_0.

Hot Spot #9 – The Relationship Between α, β, And n

✳
Mann
Section 9.1.3

The relationship between α and β is essentially that as one goes up in value, the other goes down in value. We often say that α and β are inversely related. However, the relationship is **not** like inverse variation; the product does **not** equal a constant. Furthermore, the sum of α and β is not 1. It is better to think of the two values as the opposite ends of a "teeter-totter".

Figure 9–2: Illustration Of The Relationship Between alpha and beta

As we push down on one side of the "teeter-totter", that is make α larger (heavier), the β value will become lighter (smaller). It is this inverse relationship that causes problems when we make the value of α extremely small. The value of β will then become quite large. This would be a bad situation because the power of the test of hypothesis is equal to $1 - \beta$. The power of the hypothesis test is the probability of rejecting a false null hypothesis; which is what we want to do.

One of our real concerns is then how to lower both α and β. The solution to our problem is quite easy in one sense of the word. All we have to do is increase the sample size we use in getting the sample mean or proportion. As easy as it sounds, using a larger sample size creates a number of related problems. Large sample sizes require more time, money, and energy. And at times it simply is neither practical nor possible to use a large sample.

The following problem using the binomial distribution shows how the values of α, β, and n are related. We will use $n = 10$ and $p = 0.5$ and set up the hypotheses as follows:

$$H_0: p = 0.5 \text{ and } H_a: p < 0.5$$

We will use a decision rule to reject H_0 if $x \le 3$. The probability of the type I error is determined by finding $P(x \le 3)$ when $p = 0.5$ and $n = 10$.

$$\alpha = P(x \le 3) = 0.1719$$

We need a specific value of $p < 0.5$ to find the value of β. We will choose $p = 0.2$ to illustrate the process. We need to find $P(x > 3)$ when $p = 0.2$ and $n = 10$ to find the probability of the type II error.

$$\beta = P(x > 3) = 1 - P(x \le 3) = 1 - 0.8791 = 0.1209$$

Although β is at an acceptable level (Power $= 1 - \beta = 88\%$), α is too large. We let $n = 5 \times 10 = 50$ and use a decision rule to reject H_0 if $x \le 15$; the ratio of 15 to 50 is the same as the ratio of 3 to 10.

The probability of a type I error is $P(x \leq 15)$ when $p = 0.5$ and $n = 50$.

$$\alpha = P(x \leq 15) = 0.0033$$

The value of β is found by taking $P(x > 15)$ when $p = 0.2$ and $n = 50$.

$$\beta = P(x > 15) = 1 - P(x \leq 15) = 1 - 0.9692 = 0.0308$$

We now find that we have lowered both the α and β values as a result of increasing the value of n from 10 to 50. Increasing n to even higher values will lower the values of α and β even more.

Now let us see what happens if we leave n at 10 but lower the α value. We will use the decision rule to reject H_0 if $x \leq 2$. We get the probability of the type I error by finding $P(x \leq 2)$ when $p = 0.5$ and $n = 10$.

$$\alpha = P(x \leq 2) = 0.0547$$

We find the probability of the type II error by finding the $P(x > 2)$ when $p = 0.2$ and $n = 10$.

$$\beta = P(x > 2) = 1 - P(x \leq 2) = 1 - 0.6778 = 0.3222$$

Although we lowered the value of α from 0.1719 to 0.0547, we increase the value of β from 0.1209 to 0.3222. In contrast, when we used a larger value of n we were able to lower α from 0.1719 to 0.0033 and lower β from 0.1209 to 0.0308.

The error analysis we just went through was reasonably easy to do for two reasons. The use of the HP-21S made it easy to answer the probability questions and we used a specific value of $p < 0.5$ from the alternate hypothesis. Generally speaking, we do not know a specific value for the alternate hypothesis. As a result, we seldom go through an error analysis. We simply assume that if α is kept at 5% and $n \geq 30$, the value of β will be small enough for us to have a powerful test.

Hot Spot #10 – Difference Between t Test Statistic And z Test Statistic

The formula we use for the t test statistic looks like the z test statistic we used when we substituted the sample standard deviation s for σ, the population standard deviation. This was not a problem as long as the sample size was larger than 30. When the sample size is less than or equal to 30, the sample standard deviation tends to underestimate σ. This can happen because small samples do not often include extreme values from the tails of the distribution. The effect on z, the standard score, is to make the value larger than it should be. When this happens in a hypothesis test, we increase our chance of rejecting a true null hypothesis and making a type I error.

The way we avoid making unnecessary type I errors when we do hypothesis tests with small samples is to adjust the critical value for the sample size. For example, the critical value of z for a test of hypothesis at 5% is 1.645 for a one-tailed hypothesis test on the right side of the curve. But if the sample size is reduced to 10, the critical value is increased to 1.833. As the sample sizes become smaller we continue increasing the size of the critical value. For example, the critical value becomes 2.132 when $n = 5$ and 2.920 when $n = 3$.

We get the new critical values using what is called the student's t distribution. The shape of the curve for the t distribution is bell-shaped looking like the now familiar standard normal curve used with the z table. But the curve for the t distribution is flatter at the center and the tails of the curve seem to float above the horizontal axis when compared to the standard normal curve. As the sample sizes get smaller, the curve becomes even more flattened. We say that the shape of the curve depends on what is called degrees of freedom. The degrees of freedom equal one less than the sample size. We use the letters df to represent degrees of freedom and write the following expression:

$$df = n - 1$$

The graph below shows a general comparison between the standard normal curve for z and student's t distribution curve used for t:

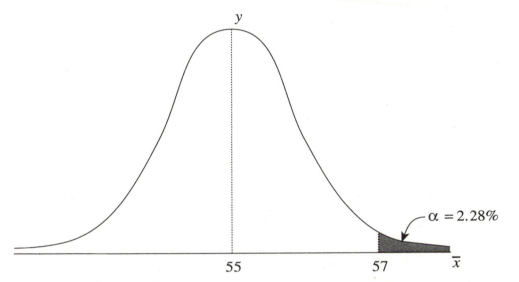

Figure 9–3:
Comparison Of
z Distribution To
t Distribution

Actually, we get a different t curve for each value of df. As a result, it is more difficult to list the critical values for the t distribution. One way this could be done is to list a new table for each value of df. But this would require at least 30 tables just to do hypothesis tests. It is now common practice to simply list the critical values for the more common significance levels. The critical value could then be found by finding the appropriate degrees of freedom in a vertical column and then moving across to the required level of α. This allows us to get by with one table of values as opposed to carrying around a small book of tables.

There is a down side to using a single t table when it comes to reporting P values for a hypothesis test. This is discussed in **Hot Spot #7**, but an easy solution to the problem of using the t table is to use a calculator like the HP-21S. It makes it easy to both get a critical value for any degrees of freedom and report the P value for a hypothesis test. The following discussion assumes that we will be using the traditional t table to find critical values.

In the traditional t table, the critical values are listed for each of the commonly accepted significance levels for each value of df below 30.

The table below shows the critical values of t for a 5% one-tailed hypothesis test as they would appear:

Critical Values of t for a 5% One-Tailed Hypothesis Test			
df	$t(0.05)$	df	$t(0.05)$
1	6.314	16	1.746
2	2.920	17	1.740
3	2.353	18	1.734
4	2.132	19	1.729
5	2.015	20	1.725
6	1.943	21	1.721
7	1.895	22	1.717
8	1.860	23	1.714
9	1.833	24	1.711
10	1.812	25	1.708
11	1.796	26	1.706
12	1.782	27	1.703
13	1.771	28	1.701
14	1.761	29	1.699
15	1.753		

Table 9–3: Critical Values Of t For A 5% One-Tailed Hypothesis Test

It is common practice to simply use the value from the z table for values of df greater than 29. Therefore, we would use $t = 1.645$ for any $df > 29$. It should again be noted that calculators like the HP-21S now allow us to find the critical value for any degree of freedom and any value of α.

If we look carefully at the two curves shown below in Figure 9–4, we can see why the critical value is larger for the t test statistic. The two curves show the critical value for a 5% one-tailed test of hypothesis. The curve used for t is at 9 degrees of freedom. We can see that we must go out further in the tail of the t curve to get 5% of the area in the tail of the curve.

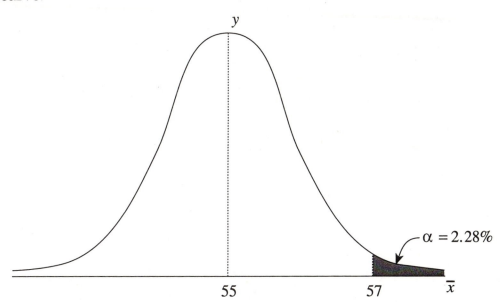

$\alpha = 2.28\%$

Figure 9–4:
5% Tails Of
z Distribution
Versus
t Distribution

The test statistic for the t distribution looks exactly like the z test statistic when we substitute s for σ.

$$t = \frac{\bar{x} - \mu}{\dfrac{s}{\sqrt{n}}}$$

We communicate that we are making the adjustment for a small sample by using the letter t for the standard score. Again, remember that we have to make the adjustment because:

1. We do not know the value of the population standard deviation.
2. The sample size is under 30.

If we know σ, the sample size is no longer a problem. When we know σ, we can use the standard normal curve and find the critical value using the z table, regardless of the sample size.

CHAPTER 9 DISCUSSION QUESTIONS

These questions may be used in your study group or simply as topics for individual reflection. Whichever you do, take time to explain verbally each topic to insure your own understanding. Since these questions are intended as topics for discussion, answers to these questions are not provided. If you find that you are not comfortable with either your answers or that your group has difficulty with the topic, take time to meet with your professor to get help.

1. What is the difference between a t and z test statistic?

2. How do you decide which side of the curve the critical value is on?

3. When is a sample considered small?

4. What is the difference between the null and the alternate hypothesis, and how do you choose the null hypothesis?

5. When do you reject the null hypothesis?

6. What is the difference between a type I and type II error?

7. Why is it important to graph the curve and locate the critical value(s) on the standard normal curve when doing a test of hypothesis?

8. What is the difference between a one-tailed and two-tailed test of hypothesis?

9. When do you use the t distribution?

10. What is the difference between α and the reported P value?

CHAPTER 9 TEST

1. A researcher wants to prove that the average cholesterol levels of athletes is less than 180. State the null and alternative hypothesis.

2. Given $H_0: \mu = 120$ and $H_a: \mu < 120$, what is the critical value of z for a hypothesis test at $\alpha = 0.05$?

3. Given the accepted view that test scores average 1,200, a sample of 50 scores gave $\bar{x} = 1,237$ and $s = 125$. Test for a significant difference at $\alpha = 0.05$.

4. A hypothesis test of $H_0: \mu = 55$ and $H_a: \mu > 55$ is done with $\alpha = 0.05$. The value of the test statistic is $z = 1.79$. Use the reported P value to complete the hypothesis test.

5. A random survey of 100 managers found that 35 of them favored a national advertising campaign. Is this a significant change from the accepted view that 40% of the managers support the campaign? Test at $\alpha = 0.05$.

Given $H_0: \mu = 55$ and $H_a: \mu \neq 55$ with $n = 75$ and $\bar{x} = 57.5$ and $s = 10$.

6. Find the critical values of t for the hypothesis test $\alpha = 0.05$.

7. Test for a significant difference at $\alpha = 0.01$.

8. Give the reported P value.

Given the following information: $\bar{x} = 1,225$, $s = 35$ and $n = 25$

9. Test $H_0: \mu = 1,200$ and $H_a: \mu > 1,200$ for significance at $\alpha = 5\%$.

10. Give the reported P value.

CHAPTER 9 TEST Questions and Answers

1. A researcher wants to prove that the average cholesterol levels of athletes is less than 180. State the null and alternative hypothesis.

Answer: $H_0: \mu = 180$ and $H_a: \mu < 180$.

2. Given $H_0: \mu = 120$ and $H_a: \mu < 120$, what is the critical value for a hypothesis test at $\alpha = 0.05$?

Answer: The critical value is -1.65.

3. Given the accepted view that test scores average 1,200, a sample of 50 scores gave $\bar{x} = 1237$ and $s = 125$. Test for a significant difference at $\alpha = 0.05$.

Answer: Step 1. $H_0: \mu = 1,200$ and $H_a: \mu \neq 1,200$.

Step 2. $n > 30$, so we use the normal distribution

Step 3. The critical value of z for a two-tailed test at $\alpha = 0.05$ is ± 1.96.

Step 4. $z = \dfrac{1,237 - 1,200}{\dfrac{125}{\sqrt{50}}} = 2.09$.

Step 5. We Reject H_0.

4. A hypothesis test of $H_0: \mu = 55$ and $H_a: \mu > 55$ is done with $\alpha = 0.05$. The value of the test statistic is $z = 1.79$. Use the reported P value to complete the hypothesis test.

Answer: $P = P(z > 1.79) = 3.67\%$. Since $P < \alpha = 0.05$, we Reject H_0.

5. A random survey of 100 managers found that 35 of them favored a national advertising campaign. Is this a significant change from the accepted view that 40% of the managers support the campaign? Test at $\alpha = 0.05$..

Answer: Step 1. $H_0: P = 0.40$ and $H_a: P \neq 0.40$.

Step 2. $n > 30$, so we use the normal distribution.

Step 3. The critical values of z for a two-tailed test at $\alpha = 0.05$ are ± 1.96.

Step 4. $z = \dfrac{\hat{p} - p}{\sqrt{\dfrac{pq}{n}}} = \dfrac{0.35 - 0.40}{\sqrt{\dfrac{(0.4)(0.6)}{100}}} = 1.02$.

Step 5. We Fail to Reject H_0.

Given $H_0: \mu = 55$ and $H_a: \mu \neq 55$ with $n = 75$ and $\bar{x} = 57.5$ and $s = 10$.

6. Find the critical values for the hypothesis test at $\alpha = 0.05$.

Answer: The critical values of z for a two-tailed test at $\alpha = 0.05$ are $z = \pm 1.96$.

7. Test for a significant difference at $\alpha = 0.01$.

Answer: Step 1. $H_0: \mu = 55$ and $H_a: \mu \neq 55$

Step 2. $n > 30$, so we use the normal distribution.

Step 3. The critical values at $\alpha = 0.01$ are $z = \pm 2.58$

Step 4. $z = \dfrac{57.5 - 55}{\dfrac{10}{\sqrt{75}}} = 2.165$

Step 5. We fail to reject H_o:

8. Give the reported P value.

Answer: Using the z table at $z = 2.17$, $\dfrac{\alpha}{2} = 0.5 - 0.485 = 0.015 \Rightarrow \alpha = 0.03$

Given the following information: $\bar{x} = 1{,}225$, $s = 35$ and $n = 25$

9. Test $H_0: \mu = 1{,}200$ and $H_a: \mu > 1{,}200$ for significance at $\alpha = 5\%$.

Answer: Step 1. $H_0: \mu = 1{,}200$ and $H_a: \mu > 1{,}200$

Step 2: $n < 30$ and we do not know σ, so we use the t distribution.

Step 3. The critical value of t for a one-tailed test at $\alpha = 0.05$ with $df = 24$ is $t = 1.711$.

Step 4. $t = \dfrac{1225 - 1200}{\dfrac{35}{\sqrt{25}}} = 3.57$

Step 5. We reject H_o

10. Give the reported P value.

Answer: Using the t table, all we can say is that $\alpha < 0.005$

Σ°

chapter ten

ESTIMATION AND HYPOTHESIS TESTING: TWO POPULATIONS

Knowledge is of two kinds.
We know a subject ourselves or
we know where we can find information upon it.

– Dr. Johnson

INTRODUCTION

In earlier chapters on hypothesis testing we used the value of a sample statistic to challenge an accepted value of a population parameter. We focused on both the mean and proportion. The goal was to show that there was a large enough difference between the value of the sample statistic and the population parameter that the difference could not be due to chance alone. When the difference was large, we argued that the value of the population parameter had changed. This approach, however, did not provide us any information about what was responsible for the change in the population parameter. In this chapter we introduce the idea of comparing two populations and use independent and dependent samples to help us determine what is responsible for the change.

When we work with **independent samples**, the two groups are often referred to as control and experimental. The control group represents what we might call the accepted value of the population parameter. The experimental group is given some treatment that we believe will change the value of the population parameter. We state a null hypothesis that the treatment used on the experimental group is not effective; that the difference in the population parameters is zero. We then compare the sample means. If the difference between the means is large enough, we argue the treatment was responsible for the change.

Interestingly enough, variance becomes a focus in this chapter. When we compare two population means, we need to be concerned about the population variances. If one of the variances is much larger than the other, the value of our test statistic may then be inflated causing us to make a type I error.

The work with independent samples is an effort to determine the effect of a treatment on the population parameter. Unfortunately the change may be a result of uncontrolled variables. One way of controlling confounding variables is to use **dependent samples** where data from the two samples is paired in a natural way. The test statistic used involves the differences of the paired data. The simplicity of this hypothesis test that the **mean difference** equals zero is complemented by the additional control of confounding variables.

In contrast, when we work with proportions, we are simply comparing two unknown population parameters. We seldom are worried about what may be responsible for the difference. We just want to know if there is a significant difference between population proportions. We state a null hypothesis that the two values are equal. If the difference between the two sample proportions is large enough we reject the null hypothesis.

CHAPTER 10 HOT SPOTS

1. **How Unequal Population Variances Can Cause Type I Errors.**
   Starts on **page 10–4.** Mann 10.1

2. **How To Decide If The Sample Data Comes From Dependent Samples.**
   Starts on **page 10–8.** Mann 10.3

3. **Why Is A Dependent Sample An Advantage In Hypothesis Testing?**
   Starts on **page 10–9.** Mann 10.3

4. **Deciding Which Test Statistic To Use For Independent Samples.**
   Starts on **page 10–13.** Mann 10.1, 10.2, 10.3

If you find other HOT SPOTS, write them down and use them as a focus of your discussions in the study group. Or you can use the HOT SPOT as the topic for a help session with your professor.

Hot Spot #1 – How Unequal Population Variances Can Cause Type I Errors

When we do a hypothesis test about the means of independent samples with n_1 or $n_2 < 30$, we must check to see if the variances of the two populations are equal. If we reject the hypothesis that population variances are equal, we then test for the null hypothesis that $\mu_1 = \mu_2$ using the degrees of freedom given by the formula:

$$df = \frac{\left(\dfrac{s_1^2}{n_1} + \dfrac{s_2^2}{n_2}\right)^2}{\dfrac{\left(\dfrac{s_1^2}{n_1}\right)^2}{n_1 - 1} + \dfrac{\left(\dfrac{s_2^2}{n_2}\right)^2}{n_2 - 1}}$$

This value is typically smaller than $n_1 + n_2 - 2$. The critical point is that the result is a more conservative hypothesis test of $\mu_1 = \mu_2$ because the critical value for the t test statistic gets larger as the degrees of freedom get smaller. The reason we need to be more conservative is illustrated in the following example.

We will consider two populations that have the same mean but very different variances, and hence different standard deviations. The graphs of the two populations are shown below.

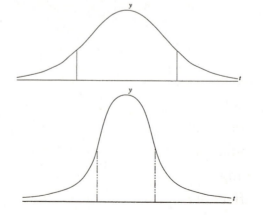

Figure 10–1:
Two t Curves
With The Same
Mean And
Differing
Standard
Deviations

Since the standard deviation is larger for the top curve (Figure 10–1), it is possible that we could get much of our data from one of the tails of the distribution. The smaller standard deviation for the bottom curve (Figure 10–1) makes it likely that most of our data will come from the center of the distribution. As such, the difference in sample means could be relatively large resulting in an incorrect decision that we reject the null hypothesis of equal population means. This would result in a type I error. The summary table below (Table 10–1) shows how this might occur. The table shows that there is a small difference in sample means, equal sample size, and a rather large difference in the sample variance.

	n	$\bar{x}$	s^2
1	15	23.08	5.4275
2	15	21.90	1.7352

Table 10–1: Summary Of Sample Sizes, Means, and Variances For Two t Curves

The standard error of $\bar{x}_1 - \bar{x}_2$ is $\sqrt{\dfrac{5.4275}{15} + \dfrac{1.7352}{15}} = 0.691$.

When we calculate the value of the t test, we get $t = \dfrac{23.08 - 21.90}{0.691} = 1.708$.

Now we need to look at the one-tailed critical value we would use if we *assume that the variances were equal*. The critical value would be $t(0.05, 28) = 1.701$ and *we would reject the null hypothesis*. However, if we *assume unequal population variances* and use the more conservative critical value for *df* = 22, we get $t(0.05, 22) = 1.717$ and *we fail to reject*. If the populations variances were unequal, the more conservative value of t using *df* for the unequal population variance test would keeps us from making a type I error. If we did get a t value larger than 1.717, we could be confident that we were rejecting the null hypothesis because there was a difference in population means.

Two different things were at work in the analysis we just finished. The first involves the effect of the variances on the numerator of the test statistic. The second point relates to the effect of the variance on the denominator of the test statistic. Before we look in detail at these two points, it is important to emphasize that when we have unequal population variances we must be very cautious. This is true not only of our work with independent samples, but also of the work we will do later in what is known as analysis of variance (ANOVA). The following analysis assumes that the population means are actually equal.

Relative to the first point, we obtained a relatively low value of the sample mean from the first distribution (on the top of Figure 10–1) because of the larger standard deviation. If the standard deviations had been equal, we would not have found so many values on the low end of one of the curves. The resulting sample means would have been closer in value and the numerator of the test statistic would be smaller. The effect would be a smaller value of the test statistic. Of course this does not take into account what happens in the denominator if the variances are equal.

The second point that we need to consider is the effect of the unequal population variances on the denominator of the test statistic. If the distributions both had the larger variance, the denominator would be larger and further reduce the size of the resulting t value. This of course, would be coupled with the reduction in the value of the numerator as explained above. The result of the two effects would be an even smaller value of the test statistic. In any case we would fail to reject the null hypothesis, unless there actually was a difference in population means.

The following calculation shows what happens to the t value if the numerator does not change and we use equal sample variances using the larger value.

$$\text{The standard error of } \bar{x}_1 - \bar{x}_2 \text{ is } \sqrt{\frac{5.4275}{15} + \frac{5.4275}{15}} = 0.851.$$

When we calculate the value of the t test, we get $t = \dfrac{23.08 - 21.90}{0.851} = 1.387$.

The result is a smaller value of t. We would fail to reject the null hypothesis in both of the hypothesis tests we did. The large variance for both populations makes it likely to get sample means that differ by a large amount and we do not find enough evidence to reject the null hypothesis of equal population means.

If both distributions had the smaller variance, the denominator would become smaller. This would appear to increase the value of the test statistic. However, it would also be unlikely to get large differences in sample means when the variances are both small. Although it is more difficult to illustrate the effect in this case, if we did find a large value for the test statistic we would again be assured that it was a result of the treatment effect and not of a difference in population variance. The following calculation shows what happens to the t value if the numerator does not change and we use equal sample variances using the smaller value.

The standard error of $\bar{x}_1 - \bar{x}_2$ is $\sqrt{\dfrac{1.7352}{15} + \dfrac{1.7352}{15}} = 0.481$.

When we calculate the value of the t test, we get $t = \dfrac{23.08 - 21.90}{0.481} = 2.453$.

We would reject the null hypothesis in both of the hypothesis tests we did. The difference in means becomes more important with the smaller variances and results in a large t value. Our hypothesis was that the population means were equal making it highly unlikely that we would get a difference of means this large. So if we did get a difference this large, it would be very significant.

Hot Spot #2 – How To Decide If The Sample Data Comes From Dependent Samples

Statistical analysis is much simpler when we collect our own data because we will know if we used dependent samples. It is when we are asked to do a hypothesis test for data that we did not collect that we have to make a decision. We must then decide if the data is from an independent sample or a dependent sample. There are two situations that indicate that we have independent samples. The first is when the sample sizes are not equal. We then know that we must have independent samples because dependent samples require the same sample size for both data sets. The second situation is when we are given a summary table that includes the means, the standard deviations, and the sample sizes. It again is very clear that we have independent samples because we do not calculate individual means and standard deviations for dependent samples.

It is when the data itself is given for the two groups that have equal sample sizes that we run into trouble. We must then search for a key word or phrase to help us out. Obviously a key word is "dependent". It makes it easy when the problem includes the information that we have been given two dependent samples. One of the key phrases to look for is "paired differences". Another phrase that may be a key is that the data represents "before and after" scores on some measuring instrument such as a test. However, there are times when there may be reasonable doubt even after looking for key phrases. We may need to look carefully at the description of how the experiment was conducted.

Dependent samples often involve the same subjects or objects. We then obtain paired data values that reflect the changes as a result of the treatment. The idea involved in a dependent sample is that by using the same subject we do not introduce confounding variables. Unfortunately there are times when it is not possible to use the same subject. For example, in studies about heart disease we often need to match patients as best we can for the confounding variables such as weight, age, life-style, etc. We then measure the treatment effect on the variable of interest, for example cholesterol or blood pressure. The paired data for the matched subjects is then treated as a dependent sample. The pairing of subjects makes it easier to find a treatment difference if it exists.

Naturally, this brings us to the point of asking, "what happens if we still can not decide if the data comes from a dependent sample?". The answer to the question is to use the independent sample hypothesis test. At worst we will make a type II error. If we instead make the assumption that the data reflects an effort to control confounding variables using the dependent sample, we could make a type I error. The possible errors are discussed further in the next **Hot Spot** that examines the effect of dependent samples on the hypothesis test.

Hot Spot #3 – Why Is A Dependent Sample An Advantage In Hypothesis Testing?

✳
*Mann
Section 10.3*

The dependent sample gives us our first look at what is called blocking. We control confounding variables by matching subjects for the variables we want to control. The result is that we are more easily able to reject a false null hypothesis. The data set that follows   (Table 10–2) illustrates the point. The example involves twenty-four males aged 25–29 selected from the Framingham Heart Study. Each of the twelve pairs of data represents a smoker and a non-smoker who were matched with regard to age and physical characteristics. The random variable measured was systolic blood pressure.

Smokers	Non-Smokers
122	114
146	134
120	114
114	116
124	138
126	110
118	112
128	116
130	132
134	126
116	108
130	116

Table 10–2:
Blood Pressures
Of Twenty-Four
Males In The
Framingham
Heart Study

We are interested in testing the hypothesis that there is a difference in mean systolic blood pressure levels for the two populations. We will do the hypothesis test using the *t* test for dependent samples. We need to find the difference between each ordered pair. The differences are listed below:

Smokers	Non-Smokers	*d*
122	114	8
146	134	12
120	114	6
114	116	-2
124	138	-14
126	110	16
118	112	6
128	116	12
130	132	-2
134	126	8
116	108	8
130	116	14

Table 10–3:
Ordered Pair
Differences Of
Blood Pressures
In Framingham
Heart Study

Using the differences as our random variable d, we find the following:

$$n = 12 \qquad \bar{d} = 6 \qquad s_d = 8.399134155$$

$$\frac{s_d}{\sqrt{12}} = 2.424621183$$

We are now ready to do the hypothesis test using the test statistic

$$t = \frac{d}{\dfrac{s_d}{\sqrt{n}}}$$

Step 1. $H_0: \mu_S = \mu_{NS}$ versus $H_a: \mu_S \neq \mu_{NS}$.

Step 2. The α level will be used at 5%.

Step 3. The two-tailed critical value of t for $df = 12 - 1$ is ± 2.201.

Step 4. $t = \dfrac{6}{2.42462} = 2.47$

Step 5. Reject H_0.

Now we will see what happens if we did not know that the data involved paired data. The following summary table gives the necessary values for the hypothesis test. We will assume that the variances for the two populations are equal.

Table 10–4: Necessary Values For Hypothesis Testing Of Twenty-Four Males In Framingham Heart Study

	n	$\bar{x}$	s^2
Smokers	12	125.67	78.4242
Non-Smokers	12	119.67	102.4242

Technically, with the variances equal and a small sample size, we should use the t test that uses the pooled standard deviation. Since the sample sizes are equal, we can use the following test statistic that will give us the same value:

$$t = \frac{\bar{x}_S - \bar{x}_{NS}}{\sqrt{\dfrac{s_S^2}{n_S} + \dfrac{s_{NS}^2}{n_{NS}}}}$$

Step 1. $H_0: \mu_S = \mu_{NS}$ versus $H_a: \mu_S \neq \mu_{NS}$.

Step 2. The α level will be used at 5%.

Step 3. The two-tailed critical value of t for $df = 24 - 2$ is ±2.074.

Step 4. $t = \dfrac{125.67 - 119.67}{\sqrt{\dfrac{78.4242}{12} + \dfrac{102.4242}{12}}} = 1.55$

Step 5. We fail to reject H_0.

Perhaps we need to review what we have just completed:

We first did a hypothesis test *assuming the data represented a dependent sample*. We found the t value of 2.47 in the critical region. So *we rejected the null hypothesis, H_0*.

We then did a hypothesis test *assuming the data represented an independent sample*. We found the t value of 1.55 that was **not** in the critical region. Thus, *we failed to reject the null hypothesis.*

The important point in all this is that when we treated the data as a dependent sample, we found a significant difference. When we assumed the data came from an independent sample we did **not** find a significant difference. Clearly, it is to our advantage to use a dependent sample whenever we can.

A second point can also be made regarding what we should do if we do not know for sure that the data is from a dependent sample. We should then go ahead and treat the data as if it were from an independent sample. If we can reject the null hypothesis under the assumption of independence, we can be confident that we would come to the same conclusion if the sample were dependent. Simply stated, the independent sample hypothesis test is a more conservative position.

When we are unable to verify that the data comes from a dependent sample and thus use the test statistic for an independent sample, we only take the chance of failing to reject a false null (a type II error). However, if we incorrectly assume that the data is from a dependent sample and use the test statistic for dependent samples, we take the chance of rejecting a true null hypothesis (a type I error).

Hot Spot #4 – Deciding Which Test Statistic To Use For Independent Samples

✳
*Mann
Sections 10.1,
10.2 And 10.3*

One of the more confusing things about this chapter is deciding on which test statistic to use for independent samples. The sense of confusion can be reduced to a very low level by looking at the decision process using a tree diagram like we used in probability.

Our first decision is to choose between a large sample and small sample case. If both of the sample sizes are 30 or more, we immediately can find the value of the z test statistic and compare it to a critical value from the z table using the values of s_1^2 and s_2^2 for the values of σ_1^2 and σ_2^2. The first branch of the tree diagram can now be drawn. It is the upper branch of the tree diagram that follows in Figure 10–2 on the next page.

If either sample is less than 30, we need to use a t test statistic and there are two choices based on whether or not the population variances are equal:

If the population variances are assumed to be equal, we use the pooled standard deviation form of the t test. It is drawn in Figure 10–2 as the top part of the lower branch.

If the population variances are not equal, we then use the test statistic with a more conservative degrees of freedom calculated from the formula shown in the diagram below. The formula is in Figure 10–2 as the bottom part of the lower branch.

The completed tree diagram is presented below in Figure 10–2. It is important to keep a visual image of the tree diagram as we look at a hypothesis test for an independent sample. The mental picture makes it easier to go through the decision process.

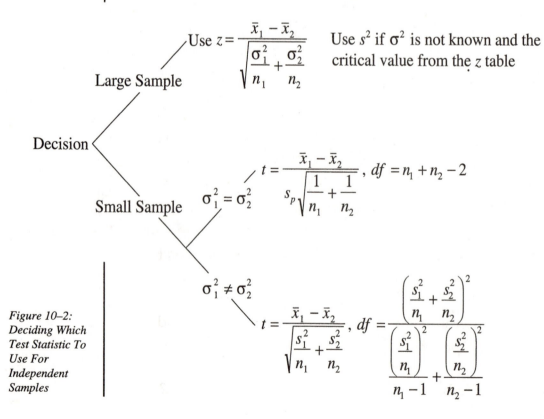

Figure 10–2: Deciding Which Test Statistic To Use For Independent Samples

Decision

Large Sample

$$\text{Use } z = \frac{\bar{x}_1 - \bar{x}_2}{\sqrt{\dfrac{\sigma_1^2}{n_1} + \dfrac{\sigma_2^2}{n_2}}}$$

Use s^2 if σ^2 is not known and the critical value from the z table

Small Sample

$\sigma_1^2 = \sigma_2^2$

$$t = \frac{\bar{x}_1 - \bar{x}_2}{s_p \sqrt{\dfrac{1}{n_1} + \dfrac{1}{n_2}}}, \quad df = n_1 + n_2 - 2$$

$\sigma_1^2 \neq \sigma_2^2$

$$t = \frac{\bar{x}_1 - \bar{x}_2}{\sqrt{\dfrac{s_1^2}{n_1} + \dfrac{s_2^2}{n_2}}}, \quad df = \frac{\left(\dfrac{s_1^2}{n_1} + \dfrac{s_2^2}{n_2}\right)^2}{\dfrac{\left(\dfrac{s_1^2}{n_1}\right)^2}{n_1 - 1} + \dfrac{\left(\dfrac{s_2^2}{n_2}\right)^2}{n_2 - 1}}$$

CHAPTER 10 DISCUSSION QUESTIONS

These questions may be used in your study group or simply as topics for individual reflection. Whichever you do, take time to explain verbally each topic to insure your own understanding. Since these questions are intended as topics for discussion, answers to these questions are not provided. If you find that you are not comfortable with either your answers or that your group has difficulty with the topic, take time to meet with your professor to get help.

1. What is the difference between an independent and dependent sample?

2. How does dependent sampling provide better control of confounding variables?

3. Why must we be concerned with the equality of population variances when we do hypothesis tests with two populations?

4. What is the test statistic for testing $\mu_1 - \mu_2$ for independent samples with equal variances?

5. Why do we use a smaller degree of freedom for a hypothesis tests about $\mu_1 - \mu_2$ when the population variances are not equal?

6. What test statistic do we use if the population variances are unequal and the size of the independent samples are small?

7. When can we use the z test statistic for a hypothesis test with independent samples from two populations?

8. When do we use the pooled estimate of p for a hypothesis test of two population proportions?

9. What is the difference between a controlled experiment and an observational experiment?

10. What is a double-blind randomized controlled experiment?

CHAPTER 10 TEST

Given:

	n	$\bar{x}$	s^2
1	21	34.5	10.38
2	30	31.3	4.67

Assume the variances are not equal.

1. What is the value of df for the hypothesis test?

2. What is the value of s_p?

3. State the formula and value of the test statistic for the hypothesis test for $H_0 : \mu_1 = \mu_2$ and $H_a : \mu_1 \neq \mu_2$ at $\alpha = 5\%$.

4. What is your conclusion?

Given: $\quad \bar{d} = 0.572 \qquad s_d = 1.58 \qquad n = 10$

5. Test $H_0 : \mu_1 = \mu_2$ and $H_a : \mu_1 \neq \mu_2$ at $\alpha = 5\%$.

6. Write the 95% confidence interval for $\mu_1 - \mu_2$.

Given: $\quad n_1 = 10, \ \bar{x}_1 = 3.15, \ s_1^2 = 1.31, \ n_2 = 15, \ \bar{x}_2 = 1.87, \ s_2^2 = 1.24$
Assume population variances are equal.

7. What is the value of the appropriate test statistic for the hypothesis test $H_0 : \mu_1 = \mu_2$ and $H_a : \mu_1 > \mu_2$ at $\alpha = 5\%$?

8. Find the 95% confidence interval for $\mu_1 - \mu_2$.

Given: $\quad \hat{p}_1 = 0.25$ and $\hat{p}_2 = 0.32$ with $n_1 = 100$ and $n_2 = 100$

9. Test the hypothesis $H_0 : p_1 = p_2$ versus $H_a : p_1 \neq p_2$ at $\alpha = 5\%$.

10. Find the 95% confidence interval for $p_1 - p_2$.

Given:

	n	$\bar{x}$	s^2
1	21	34.5	10.38
2	30	31.3	4.67

Assume the population variances are not equal.

1.   What is the value of df for the hypothesis test?

Answer: $df = \dfrac{\left(\dfrac{10.38}{21} + \dfrac{4.67}{30}\right)^2}{\dfrac{\left(\dfrac{10.38}{21}\right)^2}{20} + \dfrac{\left(\dfrac{4.67}{30}\right)^2}{29}} = 34.58 \approx 35$

2.   What is the value of s_p?

Answer: $s_p = \sqrt{\dfrac{(n_1 - 1)s_1^2 + (n_2 - 1)s_2^2}{n_1 + n_2 - 2}}$

$= \sqrt{\dfrac{20(10.38) + 29(4.67)}{49}}$

$= 2.6459$

3. State the formula and value of the test statistic for the hypothesis test for $H_0 : \mu_1 = \mu_2$ and $H_a : \mu_1 \neq \mu_2$ at $\alpha = 5\%$.

Answer: $t = \dfrac{\bar{x}_1 - \bar{x}_2}{\sqrt{\dfrac{s_1^2}{n_1} + \dfrac{s_2^2}{n_2}}} = \dfrac{34.5 - 31.3}{\sqrt{\dfrac{10.38}{21} + \dfrac{4.67}{30}}} = \dfrac{3.2}{0.8062} = 3.97$

4. What is your conclusion?

Answer: Reject H_o because $3.97 > t(35, 0.025) = 2.03$

Given: $\bar{d} = 0.572$ $\qquad s_d = 1.58$ $\qquad n = 10$

5. Test $H_0 : \mu_1 = \mu_2$ and $H_a : \mu_1 \neq \mu_2$ at $\alpha = 5\%$.

Answer: $t = \dfrac{\bar{d}}{\dfrac{s_d}{\sqrt{n}}} = \dfrac{0.572}{\dfrac{1.58}{\sqrt{10}}} = 1.14$

$t(0.025, 9) = 2.26$. Fail to reject H_0.

6. Write the 95% confidence interval for $\mu_1 - \mu_2$.

Answer: $\bar{d} - t\left(\frac{\alpha}{2}, n-1\right)\dfrac{s_d}{\sqrt{n}} \leq \mu_1 - \mu_2 \leq \bar{d} + t\left(\frac{\alpha}{2}, n-1\right)\dfrac{s_d}{\sqrt{n}}$

$0.572 - 2.26\dfrac{1.58}{\sqrt{10}} \leq \mu_1 - \mu_2 \leq 0.572 + 2.26\dfrac{1.58}{\sqrt{10}}$

$-0.557 \leq \mu_1 - \mu_2 \leq 1.701$

A Study Guide For Statistics

Given: $n_1 = 10$, $\bar{x}_1 = 3.15$, $s_1^2 = 1.31$, $n_2 = 15$, $\bar{x}_2 = 1.87$, $s_2^2 = 1.24$

Assume population variances are equal.

7. What is the value of the appropriate test statistic for the hypothesis test $H_0 : \mu_1 = \mu_2$ and $H_a : \mu_1 > \mu_2$ at $\alpha = 5\%$?

Answer: The variances can be assumed equal because $F = 1.06$ and the critical value for $F = 3.21$.

$$t = \frac{(\bar{x}_1 - \bar{x}_2) - (\mu_1 - \mu_2)}{s_p \sqrt{\dfrac{1}{n_1} + \dfrac{1}{n_2}}} = \frac{(3.15 - 1.87) - 0}{1.12578} = 1.137$$

$$s_p = \sqrt{\frac{9(1.31) + 14(1.24)}{23}} = 1.12578$$

8. Find the 95% confidence interval for $\mu_1 - \mu_2$.

Answer:

$$(3.15 - 1.87) - 1.96(1.12578) \le \mu_1 - \mu_2 \le (3.15 - 1.87) + 1.96(1.12578)$$

$$-0.93 \le \mu_1 - \mu_2 \le 3.49$$

Given: $\hat{p}_1 = 0.25$ and $\hat{p}_2 = 0.32$ with $n_1 = 100$ and $n_2 = 100$

9. Test the hypothesis $H_0: p_1 = p_2$ versus $H_a: p_1 \ne p_2$ at $\alpha = 5\%$.

Answer: $\bar{p} = \dfrac{x_1 + x_2}{n_1 + n_2} = \dfrac{25 + 32}{200} = 0.285$

$$\bar{q} = 1 - \bar{p} = 1 - 0.285 = 0.715$$

$$s_{\hat{p}_1 - \hat{p}_2} = \sqrt{\overline{pq}\left[\frac{1}{n_1} + \frac{1}{n_2}\right]}$$

$$= \sqrt{(0.285)(0.715)[0.01 + 0.01]}$$

$$= \sqrt{0.0040755}$$

$$= 0.0638396$$

$$z = \frac{\hat{p}_1 - \hat{p}_2 - (p_1 - p_2)}{\sqrt{\overline{pq}\left[\dfrac{1}{n_1} + \dfrac{1}{n_2}\right]}}$$

$$= \frac{(0.25 - 0.32) - 0}{0.0638396}$$

$$= -1.096$$

At $\alpha = 0.05$ the critical value of z is ± 1.96. Since the value of the test statistic is **not** in the critical region, we fail to reject the null hypothesis.

10. Find the 95% confidence interval for $p_1 - p_2$.

Answer:

$$\left(\hat{p}_1 - \hat{p}_2\right) - z\left(\tfrac{\alpha}{2}\right)\sqrt{\frac{\hat{p}_1\hat{q}_1}{n_1} + \frac{\hat{p}_2\hat{q}_2}{n_2}} \le p_1 - p_2 \le \left(\hat{p}_1 - \hat{p}_2\right) + z\left(\tfrac{\alpha}{2}\right)\sqrt{\frac{\hat{p}_1\hat{q}_1}{n_1} + \frac{\hat{p}_2\hat{q}_2}{n_2}}$$

$$\tfrac{1}{4} - \tfrac{32}{100} - 1.96\sqrt{\frac{\left(\tfrac{1}{4}\right)\left(\tfrac{3}{4}\right)}{100} + \frac{\left(\tfrac{32}{100}\right)\left(\tfrac{68}{100}\right)}{100}} \le p_1 - p_2 \le \tfrac{1}{4} - \tfrac{32}{100} + 1.96\sqrt{\frac{\left(\tfrac{1}{4}\right)\left(\tfrac{3}{4}\right)}{100} + \frac{\left(\tfrac{32}{100}\right)\left(\tfrac{68}{100}\right)}{100}}$$

$$-0.07 - 1.96(0.064) \le p_1 - p_2 \le -0.07 + 1.96(0.064)$$

$$-0.07 - 0.125 \le p_1 - p_2 \le -0.07 + 0.125$$

$$-0.195 \le p_1 - p_2 \le 0.055$$

$$\Sigma_\circ$$

chapter eleven

CHI–SQUARE TESTS

Don't ask what it means,
but rather how it is used.

– L. Wittgenstein

INTRODUCTION

This chapter is an extension of the hypothesis testing we did with two proportions in chapter 10. We can use the Chi-Square distribution to test hypotheses that involve more than two population proportions. We will also use the Chi-Square distribution to test hypotheses about the population variance.

The test statistic used for proportions in this chapter involves what are called **observed frequencies** and **expected frequencies** that occur when we work with **categorical data**. A **contingency table** is used to record the observed frequencies in the **cells** of the table. The value of the test statistic is found by taking the sum of the squares of the difference between the observed and expected frequencies divided by the expected frequency for each category. The test statistic has an approximate chi-square distribution, and the critical value for a given level of significance is obtained from the chi-square table. Three of the tests of hypotheses that we will look at in this chapter are called a test for goodness of fit, a test of independence, and test for homogeneity.

The **test for goodness of fit** is one of the first hypothesis tests developed; it was first published in 1900 by Karl Pearson. This test can be illustrated using an example with the coins. In tossing a fair coin there are two categories, heads and tails. In ten tosses, we expect 5 heads and 5 tails, but we may only observe 4 heads and 6 tails. A hypothesis test of the equality of the proportions for heads and tails is done using the expected values of 5 and the observed values of 4 and 6 in the test statistic. We then compare it to a critical value obtained from the χ^2 table. Of course, the goodness of fit test can be extended to more categories.

The **test of independence** is used to determine if two or more characteristics of a single population are related. A typical example of a test of independence involves data on cholesterol levels for men and women to determine if cholesterol levels are independent of gender. The notation used to find the expected proportion in each cell is similar to what we used in the probability chapter when we had independent events. For example, the expected proportion of the sample that is female and has low cholesterol, p_{FL}, is found using the product $p_F \cdot p_L$. The observed and expected frequencies are then used in the test statistic, and it is compared to a critical value from the χ^2 table.

The **homogeneity test** is similar to the test of independence, but it involves more than one population. For example, we consider five different ethnic groups and measure the number of people in each group with high cholesterol. We test the hypothesis that the five population proportions are equal. The equality of the proportions is used to obtain the expected value for each category. The observed and expected frequencies are then used to get the value of the chi-square test statistic, and it is compared to the critical value from the χ^2 table.

Although the chi-square test statistic that is used in each of the above tests looks very different from the other test statistics we have used in hypothesis testing, we still use the same five steps to determine if we should reject the null hypothesis.

CHAPTER 11 HOT SPOTS

1. z **Test For Proportions Versus Chi-Square Test.**
   Starts on **page 11–4.** Mann 11.4

2. **Reporting** P **Values For The** χ^2 **Distribution**
   Starts on **page 11–5.** Problems on **pages 11–6, 11–7.**
   Mann 11.2, 11.4

If you find other HOT SPOTS, write them down and use them as a focus of your discussions in the study group. Or you can use the HOT SPOT as the topic for a help session with your professor.

Hot Spot #1 – z Test For Proportions Versus Chi-Square Test

Earlier we used a z test statistic to test a hypothesis of equal proportions. The chi-square test of homogeneity can also be used to test a hypothesis that two population proportions are equal. The chi-square homogeneity test has the advantage of being a more general hypothesis test that allows us to test whether two or more populations are homogeneous with respect to a characteristic. The following example shows that we get exactly the same results using the two different hypothesis tests on the same data. A survey of 50 men and 50 women asked them whether they liked country music. The results are entered into the table below:

Survey of Country Music Tastes

Respondents	No	Yes	Total
Men(1)	20	30	50
Women(2)	15	35	50
Totals	35	65	100

Figure 11–1:
Survey Of
Country Music
Tastes

We want to test a null hypothesis that the population proportions are equal, against the alternate hypothesis that they are not equal, at $\alpha = 5\%$.

$$H_0: p_1 = p_2 \text{ versus } H_a: p_1 \neq p_2$$

The critical values for the two-tailed test are ±1.96, and we will test the hypothesis using the test statistic

$$z = \frac{\hat{p}_1 - \hat{p}_2}{\sqrt{\hat{p}\hat{q}\left[\frac{1}{n_1} + \frac{1}{n_2}\right]}}$$

$$\hat{p}_1 = \frac{30}{50} \qquad\qquad \hat{p}_2 = \frac{35}{50}$$

$$\hat{p} = \frac{30+35}{100} = \frac{65}{100} \qquad\qquad \hat{q} = \frac{35}{100}$$

$$z = \frac{\frac{30}{50} - \frac{35}{50}}{\sqrt{\frac{65}{100}\left(\frac{35}{100}\right)}\sqrt{\frac{1}{50} + \frac{1}{50}}} = -1.048$$

The value of z is between the critical values of -1.96 and 1.96; hence, we fail to reject the null hypothesis. Using the HP-21S, the reported P value is 0.147, which we must double because of the two-tailed test. The value would be at 29.4%, which is much too high.

Now we will do the hypothesis test using the chi-square test of homogeneity. We first need to find the expected values for each cell of the table, and then calculate X^2.

$$E_{1,1} = \frac{50(35)}{100} = 17.5 \qquad E_{1,2} = \frac{50(65)}{100} = 32.5$$

$$E_{2,1} = \frac{50(35)}{100} = 17.5 \qquad E_{2,2} = \frac{50(65)}{100} = 32.5$$

$$X^2 = \frac{(20-17.5)^2}{17.5} + \frac{(30--32.5)^2}{32.5} + \frac{(15-17.5)^2}{17.5} + \frac{(35-32.5)^2}{32.5}$$

$$X^2 = 0.357 + 0.192 + 0.357 + 0.192 = 1.098$$

The critical value of χ^2 at one degree of freedom is 3.84. The value of X^2 is less than the critical value; hence, we again fail to reject the null hypothesis. Using the HP-21S the reported P value for α is 29.4%. This is exactly the same value we found using the z test. The advantage of the z test is that one can test one tailed hypotheses. The advantage of the χ^2 test is that it can be generalized to more than two proportions.

Hot Spot #2– Reporting P Values For The χ^2 Distribution

The table used to get critical values for the chi-square distribution is similar to the t table. The critical values listed for each value of α are found by first finding the degrees of freedom. However, there is a very important difference in the tables when we find the critical value for the left side of the curve.

❋
Mann
Sections
11.2 And 11.4

The chi-square distribution unlike the z and t distributions, is not symmetrical. Because of the non-symmetry we list the critical values for each of the two tails of the curve. It is no longer possible to simply use a $\pm$ sign in front of the value we found for the right side of the curve. When we look at the chi-square table we find a heading that gives areas to the right of the critical value for both sides of the curve. A typical heading and row of critical values for $df = 9$ follows:

df	$\chi^2(0.99)$	$\chi^2(0.975)$	$\chi^2(0.95)$	$\chi^2(0.90)$	$\chi^2(0.10)$	$\chi^2(0.05)$	$\chi^2(0.01)$
9	2.088	2.700	3.325	4.168	14.684	16.919	21.666

The values in the table should make it very clear that we cannot simply use the $\pm$ symbol to get the value for the left side of the curve. For example, at $\alpha = 0.05$ the critical value of χ^2 for the right side is 16.919, but the critical value for the left side is 3.325.

When we do a hypothesis test using the χ^2 table we find the critical value or values consistent with whether we have a one-tailed or two-tailed test. We, of course, also need the significance level. It is when we decide to do a hypothesis test using a reported P value that we run into the same difficulty we had with reporting P values with the t distribution.

Before we look at how we handle the problem with the chi-square table, it is again important to understand that we can avoid the problem using a calculator like the HP-21S. Once we have the degrees of freedom entered into the calculator, we can find the area to the right of our observed value by pressing one key. For that matter, we can get critical values for a hypothesis test just as easily. However, the remainder of the discussion assumes that we will use the traditional χ^2 table to report the P value associated with a hypothesis test.

As an example, we will do a hypothesis test on $H_0: \sigma^2 = 1.25$ and $H_a: \sigma^2 > 1.25$ at $\alpha = 0.05$ with $n = 10$. The sample gave $s^2 = 2.20$. When we calculate the test statistic we get the following:

$$\chi^2 = \frac{(n-1)s^2}{\sigma^2} = \frac{(9)(2.2)}{1.25} = 15.84$$

We then compare the calculated value to the table entries for $df = 9$. The calculated value is between 14.684 at $\alpha = 10\%$ and 16.919 at $\alpha = 5\%$. We would say that the hypothesis test was significant at a value between 5% and 10%. (Using the HP-21S, we find the exact value of $\alpha = 0.070$.)

Hot Spot #2 Sample Problem and Solution: Find the reported P value for the following hypothesis test on $H_0: \sigma^2 = 0.25$ and $H_a: \sigma^2 \neq 0.25$ with $s^2 = 0.5$ and $n = 10$.

Solution: $\chi^2 = \dfrac{(9)(0.5)}{0.25} = 18$ and $df = 10 - 1 = 9$.

18 is between 16.919 at $\frac{\alpha}{2} = 5\%$ and 19.023 at $\frac{\alpha}{2} = 2\frac{1}{2}\%$. The reported value of P is between 5% and 10%.

Sample Problem and Answer: Find the reported P value for a one-tailed hypothesis test on the left side of the curve if $\chi^2 = 5.22$ and $n = 15$.

Solution: We use $df = 15 - 1 = 14$.

5.22 is between 4.660 at $\alpha = 1\%$ and 5.629 at $\alpha = 2\frac{1}{2}\%$. The reported P value is between 1% and $2\frac{1}{2}\%$.

Sample Problem and Answer: Find the reported P value for a two-tailed hypothesis test if $\chi^2 = 50.475$ and $n = 30$.

Solution: We use $df = 30 - 1 = 29$.

50.475 is between 49.588 at $\frac{\alpha}{2} = 1\%$ and 52.336 at $\frac{\alpha}{2} = 0.5\%$. The reported P value is between 1% and 2%.

CHAPTER 11 DISCUSSION QUESTIONS

These questions may be used in your study group or simply as topics for individual reflection. Whichever you do, take time to explain verbally each topic to insure your own understanding. Since these questions are intended as topics for discussion, answers to these questions are not provided. If you find that your are not comfortable with either your answers or that your group has difficulty with the topic, take time to meet with your professor to get help.

1. What is the difference between a test for goodness-of-fit and a test of independence?

2. What is the difference between a test for goodness-of-fit and a homogeneity test?

3. What is the difference between a test of independence and a homogeneity test?

4. What is the minimum expected frequency for any cell?

5. How do the row totals and column totals help determine if one is doing a test of independence or a test of homogeneity?

6. How does one find the expected frequencies for the cells for tests of independence and homogeneity tests?

7. What is the test statistic for a test of hypothesis about σ^2?

8. What is a contingency table?

9. How does one find degrees of freedom for the chi-square test?

10. How does the shape of the chi-square curve for $df = 1$ or $df = 2$ differ from the shape of the distribution for other values of degrees of freedom?

CHAPTER 11 TEST

To determine if three comparable restaurants are equally popular, a survey was done during a Friday evening to find the number of people at each restaurant. The results were 28 in restaurant 1, 52 in restaurant 2, and 49 in restaurant 3.

1. State the null and alternate hypothesis for a test of goodness of fit.

2. Find E, the expected frequency, for each cell.

3. Find the value of X^2 and state your conclusion.

At one of the restaurants during a weekend, the manager surveyed the customers regarding their preference of music:

	Easy Listening	Hard Rock	Country Western	Total
Men	17	19	13	49
Women	16	12	23	51
Totals	33	31	36	100

4. State the null and alternate hypothesis for a test of independence.

5. Find E, the expected frequency, for each cell.

6. Find the X^2 value and state your conclusion at $\alpha = 10\%$.

A marketing consultant is hired by the three restaurants to determine the age of the customers. The results are:

Restaurant:	1	2	3	Total
Under 30	10	30	12	52
30 and Over	18	22	37	77
Totals	28	52	49	129

7. State the null and alternate hypothesis for the homogeneity test.

8. Find E, the expected frequency, for each cell.

9. Find the X^2 value and state your conclusion at $\alpha = 1\%$.

10. Test the hypothesis $H_o: \sigma^2 = \frac{1}{4}$ versus $H_a: \sigma^2 > \frac{1}{4}$ at $\alpha = 5\%$; where $s^2 = \frac{1}{2}$ and $n = 20$.

A Study Guide For Statistics

CHAPTER 11 TEST Questions and Answers

To determine if three comparable restaurants are equally popular, a survey was done during a Friday evening to find the number of people at each restaurant. The results were 28 in restaurant 1, 52 in restaurant 2, and 49 in restaurant 3.

1. State the null and alternate hypothesis for a test of goodness of fit.

Answer: $H_0: p_1 = p_2 = p_3$ $H_a: H_0$ is not true.

2. Find E, the expected frequency, for each cell.

Answer: $E = np = 129\left(\frac{1}{3}\right) = 43$.

3. Find the value of X^2 and state your conclusion.

Answer: $X^2 = \dfrac{(28-43)^2}{43} + \dfrac{(52-43)^2}{43} + \dfrac{(49-43)^2}{43}$

$= \dfrac{225}{43} + \dfrac{81}{43} + \dfrac{36}{43} = \dfrac{342}{43} = 7.953.$

The value of X^2 is larger than the critical value of 5.991, so we reject the null hypothesis that the restaurants are equally popular.

At one of the restaurants during a weekend, the manager surveyed the customers regarding their preference of music:

	Easy Listening	Hard Rock	Country Western	Total
Men	17	19	13	49
Women	16	12	23	51
Totals	33	31	36	100

4. State the null and alternate hypothesis for a test of independence.

Answer: H_0: Gender and music preference are independent.

H_a: Gender and music preference are related.

5. Find E, the expected frequency, for each cell.

Answer:

	Easy Listening	Hard Rock	Country Western
Men	$\dfrac{49(33)}{100} = 16.17$	$\dfrac{49(31)}{100} = 15.19$	$\dfrac{49(36)}{100} = 17.64$
Women	$\dfrac{51(33)}{100} = 16.83$	$\dfrac{51(31)}{100} = 15.81$	$\dfrac{51(36)}{100} = 18.36$

6. Find the X^2 value and state your conclusion at $\alpha = 10\%$.

Answer: $X^2 = \dfrac{(17-16.17)^2}{16.17} + \dfrac{(19-15.19)^2}{15.19} + \dfrac{(13-17.64)^2}{17.64} +$

$\dfrac{(16-16.83)^2}{16.83} + \dfrac{(12-15.81)^2}{15.81} + \dfrac{(23-18.36)^2}{18.36}$

$X^2 = 0.04 + 0.96 + 1.22 + 0.04 + 0.92 + 1.17 = 4.350.$

The value of X^2 is smaller than the critical value of $\chi^2(0.10,(3-1)(2-1)) = 4.605$, so we fail to reject the null hypothesis that gender and music preference are independent.

A marketing consultant is hired by the three restaurants to determine the age of the customers. The results are:

Restaurant:	1	2	3	Total
Under 30	10	30	12	52
30 and Over	18	22	37	77
Totals	28	52	49	129

7. State the null and alternate hypothesis for the homogeneity test.

Answer: H_0: The three populations are homogeneous with respect to the proportion of customers under 30.
H_a: The three populations are not homogeneous with respect to this characteristic.

8. Find E, the expected frequency, for each cell.

Answer: $E_{1,1} = \dfrac{52(28)}{129} = 11.287 \quad E_{2,1} = \dfrac{77(28)}{129} = 16.713$

$E_{1,2} = \dfrac{52(52)}{129} = 20.961 \quad E_{2,2} = \dfrac{77(52)}{129} = 31.039$

$E_{1,3} = \dfrac{52(49)}{129} = 19.752 \quad E_{2,3} = \dfrac{77(49)}{129} = 29.248$

9.   Find the X^2 value and state your conclusion at $\alpha = 1\%$.

Answer: $X^2 = \dfrac{(10-11.287)^2}{11.287} + \dfrac{(30-20.961)^2}{20.961} + \dfrac{(12-19.752)^2}{19.752} +$

$\dfrac{(18-16.713)^2}{16.713} + \dfrac{(22-31.039)^2}{31.039} + \dfrac{(37-29.248)^2}{29.248}$

$X^2 = 0.147 + 3.898 + 3.042 + 0.099 + 2.632 + 2.055 = 11.873.$

The degrees of freedom are $df = (3-1)(2-1) = 2$.

$$\chi^2(0.01, 2) = 9.210$$

The value of X^2 is larger than the critical value, so we reject the null hypothesis and conclude that the three populations are not homogeneous with respect to the proportion of customers under the age of 30.

10.   Test the hypothesis $H_o : \sigma^2 = \tfrac{1}{4}$ versus $H_a : \sigma^2 > \tfrac{1}{4}$ at $\alpha = 5\%$;

where $s^2 = \tfrac{1}{2}$ and $n = 20$.

Answer: $\chi^2(0.05, 19) = 30.144$ $\chi^2 = \dfrac{(19)(\frac{1}{2})}{\frac{1}{4}} = 38$ Reject H_o

$\sum \circ$

chapter twelve

ANALYSIS OF VARIANCE

Variability is the law of life.
> – Sir William Osler

INTRODUCTION

This chapter is a natural extension of the hypothesis tests we have done using inferential statistics. The use of the standard score as a test statistic with single samples in chapter 9 was focused on the difference of the sample mean and population mean. However, the value of the test statistic was really dependent on the value of the standard deviation, and hence the variance. As we extended the hypothesis tests to small samples in chapter 9 and independent samples in chapter 10, the focus on means was subtly changed. When the sample or samples were small, we had to consider the effect that variance had on the value of the test statistic. In essence our hypothesis tests involved an analysis of variance. In fact, all of the hypothesis tests we have done up to this point are special cases of a more general analysis that focuses on variance. The generalization is called analysis of variance, or **ANOVA**.

The advantage of using ANOVA is that it allows us to extend the experimental design of a control group versus a treatment group to a design that considers different levels of treatment. With ANOVA, we test a null hypothesis that all of the population means are equal against an alternate hypothesis that there is at least one mean not equal to the others. It might, at first, seem reasonable to assume that ANOVA involves repeated experiments that compare all of the treatments, two at a time. However, the repeated experiment design would create a problem with the size of the type I error when we reject H_0. If we use repeated experiments, the probability of rejecting at least one null hypothesis when the population means are equal would become unacceptably large. The ANOVA design actually involves all levels of treatment as part of one experiment.

We find the sample mean and variance for each of the k levels of treatment. Using these values, we obtain two different estimates of the population variance. The first estimate of the population variance is obtained by finding the sample variance of the k sample means from the **grand mean**. This **between sample estimate of** σ^2 is referred to as the variance between the means. The second estimate of the population variance is found by using a weighted average of the sample variances. This **within sample estimate of** σ^2 is called the variance within the means.

The two estimates of the population variance are then compared using the F test, with F equal to **Mean Squares Between** divided by the **Mean Squares Within**. This calculated value of F is then compared to a critical value of F at $n - k$ and $k - 1$ degrees of freedom to determine if we should reject the null hypothesis. When there is no treatment effect, the ratio should be close to 1. If there is a treatment effect, the observed value of F will be greater than the critical value. We then reject the null hypothesis and argue that at least one of the treatments was effective.

The F test assumes normal populationswith equal variance, and independent samples. Analysis of variance is usually done using an **ANOVA table** that is generated by a computer. The analysis is sensitive to inequality of variance (**heteroscedasticity**) when the sample sizes are small and unequal and care should be used in interpreting the results.

CHAPTER 13 HOT SPOTS

1. **The *t* Test Versus ANOVA**

   Starts on **page 12–4.** Mann 12.2

2. **The Importance Of The ANOVA Table**

   Starts on **page 12–6.** Mann 12.2

If you find other HOT SPOTS, write them down and use them as a focus of your discussions in the study group. Or you can use the HOT SPOT as the topic of a help session with your professor.

Hot Spot #1 – The t Test Versus ANOVA

When we have two treatment levels, the t test used with independent samples is just a special case of ANOVA. The following example illustrates the point:

$$A = \text{Treatment group} \qquad B = \text{Control group}$$
$$\bar{x}_A = 57.3 \qquad\qquad \bar{x}_B = 54.7$$
$$s_A^2 = 12.4 \qquad\qquad s_B^2 = 12.6$$
$$m_A = 25 \qquad\qquad m_B = 25$$

Test $H_0: \mu_A = \mu_B$ and $H_a: \mu_A \neq \mu_B$ at $\alpha = 5\%$.

Using the t test for independent samples with variances equal and the short cut for sample sizes equal,

$$t = \frac{57.3 - 54.7}{\sqrt{\dfrac{12.4}{25} + \dfrac{12.6}{25}}} = \frac{2.6}{\sqrt{\dfrac{25}{25}}} = 2.6$$

The observed value of $t = 2.6$ is in the rejection region determined by the critical values of $t\left(\frac{0.05}{2}, 48\right) = \pm 2.0106$. Therefore, we reject H_0 and state that there is a significant difference in population means.

Using the F test for ANOVA we do the following analysis:

$$\bar{\bar{x}} = \frac{57.3 + 54.7}{2} = 56$$

$$s_{Between}^2 = \frac{(57.3 - 56)^2 + (54.7 - 56)^2}{(2 - 1)} = 3.38$$

$$s_{Within}^2 = \frac{12.4 + 12.6}{2} = 12.5$$

Using $F = \dfrac{ns_B^2}{s_W^2}$, we get $F = \dfrac{(3.38)(25)}{12.5} = 6.76$

The observed value of $F = 6.76$ is in the rejection region

determined by the critical value of $F(0.05, 1, 48) = 4.0427$. Therefore, we reject the hypothesis that the population means are equal.

We now compare the results of the two hypothesis tests:

t Test	ANOVA
$\alpha = 0.05$	$\alpha = 0.05$
$t = 2.6$	$F = 6.76$
$t\left(\frac{0.05}{2}, 48\right) = \pm 2.0106$	$F(0.05, 1, 48) = 4.0427$

Table 12–1: Comparison of Two Hypothesis Tests: t Test Versus ANOVA

The point of the above comparison is that if we square both the observed t value and the critical value of t, we get the values of F we have in the ANOVA column. The F test for two treatment levels is simply the square of the t test for independent samples. The critical values are related using the formula:

$$t^2\left(\frac{\alpha}{2}, n-2\right) = F(\alpha, 1, n-2)$$

Thus, the t test we used with two independent samples with population variances equal, is a special case of ANOVA with $k = 2$ treatment levels. There are three advantages to using ANOVA. The first is that we then have an experimental design that will work with more than two treatment levels. The second advantage is that the t test is very sensitive to errors that occur in rounding off values to get the mean and standard deviation. The third advantage is in being able to use the F test rather than having to know and remember both the t and F test statistics.

The above comparison between the t test and ANOVA was done using two groups of equal sample size. This was done to simplify the comparison. However, ANOVA can be used with treatment levels involving groups of unequal sample size.

The only restrictions in using ANOVA are that the populations are normal, the population variances are equal, and the random samples are independent. If these assumptions are violated, we can use the Kruskal–Wallis test from what is called nonparametric statistics.

✳
Mann
Section 12.2

Hot Spot #2 – The Importance Of The ANOVA Table

There is an obvious advantage to using a computer to do the number crunching that is involved with large samples. As the number of treatment levels increase, it is imperative that the computer be used to avoid errors in the analysis of the data. Most of the computer outputs for analysis of variance are done with the ANOVA table. As such, it is worthwhile to learn how to read an ANOVA table. However, there is an even more important reason for becoming familiar with the ANOVA table.

Many of the statistical software packages used on the computer treat bivariate data analysis (Regression and Correlation) as a special case of what is called multivariate statistical analysis. The output of these programs includes the t tests for the null hypotheses that the coefficients of the independent variables are each zero. The computer output reports the level of significance for each t test and includes the ANOVA table for the null hypothesis that the population multiple correlation coefficient is zero.

The ANOVA table becomes a common structure for both Analysis of Variance and Multivariate Data Analysis that includes Multiple Regression and Multiple Correlation. Because of this it is important to

1. become familiar with the ANOVA table,
2. become comfortable with the terminology,
3. know how to interpret the results.

CHAPTER 12 DISCUSSION QUESTIONS

These questions may be used in your study group or simply as topics for individual reflection. Whichever you do, take time to explain verbally each topic to insure your own understanding. Since these questions are intended as topics for discussion, answers to these questions are not provided. If you find that you are not comfortable with either your answers or that your group has difficulty with the topic, take time to meet with your professor to get help.

1. What do the letters ANOVA represent?

2. What is the null hypothesis using ANOVA?

3. Why do we use ANOVA for k treatments rather than doing all possible comparisons two at a time?

4. What assumptions are made when using ANOVA?

5. How critical are the assumptions when using ANOVA?

6. What is meant by heteroscedasticity?

7. What is the difference in the phrases "estimate between" and "estimate within"?

8. How do we avoid round off errors in ANOVA?

9. What is an ANOVA table?

10. What is the relationship between the following phrases?
    Between   Within

CHAPTER 12 TEST

Given the following information, use ANOVA:

A=Treatment 1	B=Treatment 2	C=Treatment 3
$\bar{x}_A = 56.1$	$\bar{x}_B = 54.9$	$\bar{x}_C = 52.4$
$s_A^2 = 10.8$	$s_B^2 = 10.2$	$s_C^2 = 10.5$
$m_A = 20$	$m_B = 20$	$m_C = 20$

1. State the null and alternate hypothesis.

2. Find $\bar{\bar{x}}$.

3. Find s_B^2.

4. Find s_W^2.

5. What is the observed value of F?

6. What is the critical value of F? State your conclusion.

Given the following ANOVA table:

Source	SS	df	MS	F
Between Samples	112	2		
Within Samples	238	27		

7. Find MSB.

8. Find MSW

9. What is the value of the F statistic?

10. What is the critical F value at $\alpha = 5\%$? State your conclusion.

CHAPTER 12 TEST Questions and Answers

Given the following information, use ANOVA:

A=Treatment 1	B=Treatment 2	C=Treatment 3
$\bar{x}_A = 56.1$	$\bar{x}_B = 54.9$	$\bar{x}_C = 52.4$
$s_A^2 = 10.8$	$s_B^2 = 10.2$	$s_C^2 = 10.5$
$m_A = 20$	$m_B = 20$	$m_C = 20$

1.   State the null and alternate hypothesis.

Answer: $H_0: \mu_A = \mu_B = \mu_C$ and H_a: not all means equal.

2.   Find $\bar{\bar{x}}$.

Answer: $\bar{\bar{x}} = \dfrac{56.1 + 54.9 + 52.4}{3} = 54.47$.

3.   Find s_B^2.

Answer: $s_B^2 = \dfrac{(56.1 - 54.47)^2 + (54.9 - 54.47)^2 + (52.4 - 54.47)^2}{3 - 1}$

$s_B^2 = \dfrac{2.66 + 0.18 + 4.28}{2} = \dfrac{7.12}{2} = 3.56$.

4.   Find s_W^2.

Answer: $s_W^2 = \dfrac{10.8 + 10.2 + 10.5}{3} = \dfrac{31.5}{3} = 10.5$

5.   What is the observed value of F?

Answer: $F = \dfrac{20(3.56)}{10.5} = 6.78$.

Analysis Of Variance

6. What is the critical value of F? State your conclusion.

Answer: $F(0.05, 3-1, 60-3) = F(0.05, 2, 57) = 3.159$.

Conclusion: Reject H_0.

Given the following ANOVA table:

Source	SS	df	MS	F
Between Samples	112	2		
Within Samples	238	27		

7. Find *MSTR*.

Answer: $MSB = \dfrac{SSB}{k-1} = \dfrac{112}{2} = 5$

8. Find *MSW*

Answer: $MSW = \dfrac{SSW}{n-k} = \dfrac{238}{27} = 8.81$

9. What is the value of the F statistic?

Answer: $F = \dfrac{MSB}{MSW} = \dfrac{56}{8.81} = 6.6$

10. What is the critical F value at $\alpha = 5\%$? State your conclusion.

Answer: $F(0.05, 3-1, 30-3) = F(0.05, 2, 27) = 3.35$.

Conclusion: Reject H_0.

$\Sigma\circ$

chapter thirteen

SIMPLE LINEAR REGRESSION

But of all these principles,
least squares is the most simple:
by the others we would be
led into the most complicated calculations.

– K.F. Gauss (1809)

INTRODUCTION

All of our work thus far has involved a single variable. However, many of the problems that we work with in statistics involve two variables. The sets of data for these problems contain ordered pairs. The first value of the ordered pair is called the **independent variable**, and the second value of the ordered pair is called the **dependent variable**. We often refer to our work with two variables as bivariate data analysis. We divide the analysis into two areas, **regression analysis** and **correlation analysis**.

One of the first things we do with bivariate data is plot the ordered pairs on a rectangular coordinate system. In many cases the graph looks like the points are clustered around a straight line. We then say that there appears to be a linear relationship between the two variables. Our first problem is to find the equation of the line that best explains the linear relationship. We use the familiar **slope-intercept form**, $y = a + bx$, of an equation for a line from algebra. The equations we use for a and b involve the Σ notation that we worked with in chapter 3. In particular, we will use the sums, Σx, Σy, Σx^2, Σy^2, and Σxy.

In general, it is impossible to find a line that all the points will fall on. So we focus on the distances (**errors**) between the points and the line we draw. We want the line that minimizes the squares of the distances using the **least squares method**. This line is called the **line of best fit** or the **regression line**. We are then able to use the equation of this line to predict the approximate value of one variable when we know the value of the other variable. The approximate value we get from the regression line is a point estimate. However, we can also create an interval estimate for the value using a confidence interval.

If few of the points we plot are on the line, we say that the linear relationship is weak with a value of b close to 0. We may then want to test a hypothesis that there is no linear relationship between the two variables. This would be the same as saying that the true value, B, of the slope of **the population regression line** is 0. We can again use a t test to test the null hypothesis, $H_o : B = 0$ against either a one-tailed or two-tailed alternate hypothesis.

If most of the points we plot are on the line or close to the line, we argue that there is a strong linear relationship between the two variables. We then feel good about using the equation for the regression line to predict values for one of the variables. We use what is called the linear **coefficient of determination** to measure the strength of the linear relationship. The coefficient of determination is the fraction of **explained variation** to the **total variation** in y.

The **correlation coefficient**, r, is the square root of the coefficient of determination. The correlation coefficient is a value between 0 and 1 for a line with a positive slope, and between -1 and 0 for a line with a negative slope. When all the points fall on the line, $r = 1$ or $r = -1$.

CHAPTER 13 HOT SPOTS

1. **How To Write The Equation Of A Line In Slope-Intercept Form.**
   Starts on **page 13–4.** Problems on **pages 13–4, 13–5, 13–6.**
   Mann 13.1.2, 13.2

2. **Estimating The Slope And Intercept Of The Regression Line.**
   Starts on **page 13–6.** Problems on **pages 13–9, 13–10.**
   Mann 13.2.1

3. **The Effect Of Errors On The Regression Line And The Correlation Coefficient.**
   Starts on **page 13–11.** Mann 13.2.2

4. **Using A** *t* **Test Statistic For A Hypothesis Test Of** $H_o : B = 0$ **Versus** $H_a : B \neq 0.$
   Starts on **page 13–13.** Problems on **pages 13–14, 13–15.**
   Mann 13.5.3

5. **The Danger Of Using The Regression Line To Predict Values Of** *y* **For** *x* **Outside The Interval Used To Obtain The Line.**
   Starts on **page 13–15.** Mann 13.2.3, 13.2.5, 13.9

If you find other HOT SPOTS, write them down and use them as a focus of your discussion in the study group. Or you can use the HOT SPOTS as the topic for a help session with your professor.

Hot Spot #1 – How To Write The Equation Of A Line In Slope-Intercept Form

In an algebra class we learn that the equation of a line can be written in the form $y = a + bx$, where b is the slope of the line and a is the y intercept. This form is known as the slope-intercept form for the equation of a line. Once we know the slope–intercept form of the equation, we can use the y intercept and slope of the line to graph the line.

The slope-intercept form we use in regression analysis is very much like going backwards from what we do in an algebra class. In algebra, we change an equation to the slope-intercept form to help us graph the line. In statistics, we have a set of ordered pairs and our problem is to find the equation of the line that best fits the bivariate data. We can use a scattergram to plot the data and draw what we believe to be a good estimate of the line of best fit. We can then use the line to find approximate values for the slope and intercept. Once we have these values, we can write the equation of the line given by our estimates.

In regression analysis, we only have a set of ordered pairs, and our problem is to find the equation of the line that best fits the bivariate data. The next **Hot Spot** shows how we can "eyeball" our best estimate for the line that fits the data. We can then use the line to find approximate values for the slope and intercept. Once we have these values, we can write the equation of the line given by our estimates.

The following example gives an illustration of what we do in an algebra class when we use the slope-intercept form of the equation to graph a line.

Hot Spot #1 Sample Problem and Solution: Given the equation $2x + 3y = 4$, use the slope-intercept form to graph the related line.

Solution: Step 1. We solve for y to get the slope-intercept form as follows:

$$y = -\tfrac{2}{3}x + \tfrac{4}{3}$$

Step 2. The coefficient of x is the slope $b = -\tfrac{2}{3}$. The constant is the y intercept $a = \tfrac{4}{3}$.

Step 3. To graph the line, we first locate the y intercept at $\left(0, \tfrac{4}{3}\right)$ and then use the slope to get a second point by starting at the y intercept and going 3 units to the right, followed by 2 units down. The graph in Figure 13–1 shows how it is done.

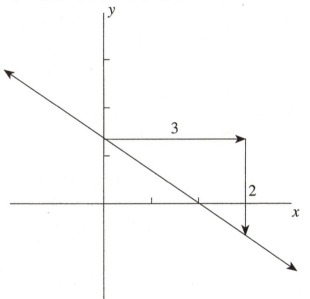

Figure 13–1:
Function Graph
$y = -\tfrac{2}{3}x + \tfrac{4}{3}$

Sample Problem and Answer: Given the equation $3x - 2y = 8$, write the equation in slope-intercept and graph the line.

Answer: $y = \tfrac{3}{2}x - 4$ and $b = \tfrac{3}{2} = 1.5$ and $a = -4$. The graph is shown below, in Figure 13–2.

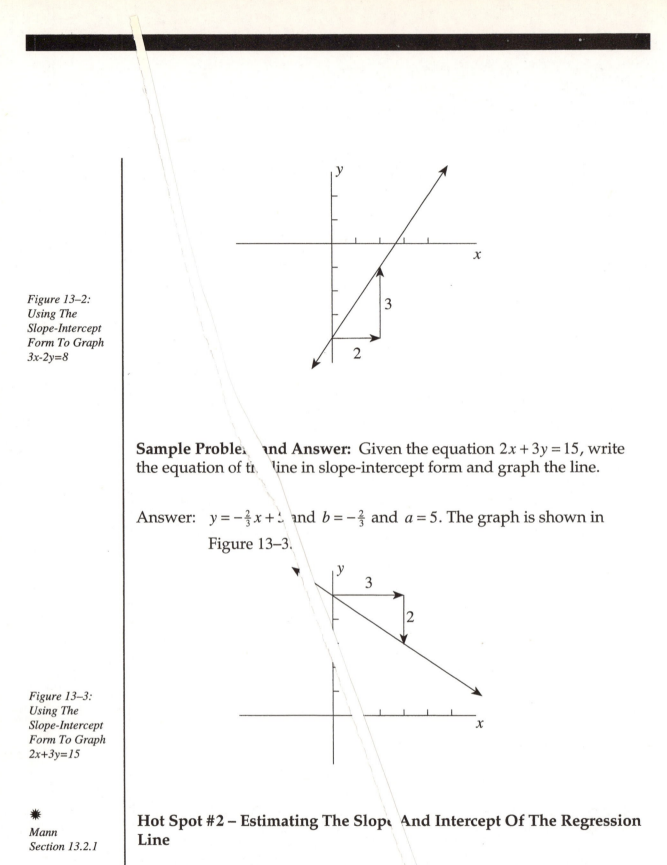

Figure 13–2:
Using The
Slope-Intercept
Form To Graph
3x-2y=8

Sample Problem and Answer: Given the equation $2x + 3y = 15$, write the equation of the line in slope-intercept form and graph the line.

Answer: $y = -\frac{2}{3}x + 5$ and $b = -\frac{2}{3}$ and $a = 5$. The graph is shown in Figure 13–3.

Figure 13–3:
Using The
Slope-Intercept
Form To Graph
2x+3y=15

Hot Spot #2 – Estimating The Slope And Intercept Of The Regression Line

When we have a set of bivariate data, we can plot the points and get what we call a scatter diagram. We then draw a line through the data that to the best of our ability, contains as many points as possible, and minimizes the vertical distance from the points to the line for the points not on the line.

It may take some practice to consistently draw a line that is close to the actual line of least squares. Sometimes it helps to lightly draw several lines through the scatter diagram to get an idea of which one is best. Even then, there may be more than one line that seems acceptable. Just remember, we are only trying to get approximate values for the slope and intercept. If there are two equally reasonable lines, we go ahead and take one at random (perhaps flipping a coin if necessary to choose). The approximate values for the slope and y intercept that we get from the line will be our estimate for the values of the regression line we get using the least squares criterion.

The following scatter diagram (Figure 13–4) will be used as an example. Two lines have been drawn through the bivariate data set. The solid line will be used to estimate the values for the slope and y intercept.

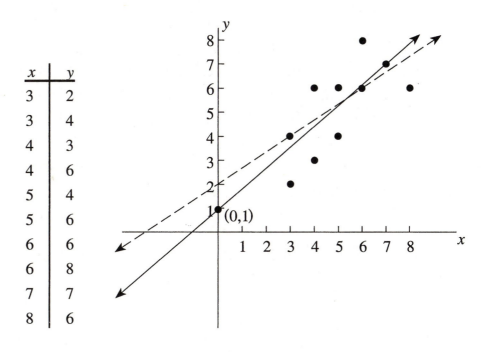

x	y
3	2
3	4
4	3
4	6
5	4
5	6
6	6
6	8
7	7
8	6

Figure 13-4: Drawing A Regression Line Through A Scatter Diagram

We need to extend the line so it cuts the vertical axis to find the value of the y intercept. In this example the line cuts the vertical axis at 1. We must be careful that both the x and y axis start at the origin without breaks. The modified form of the above graph (Figure 13–4) shows what can happen if we are not careful.

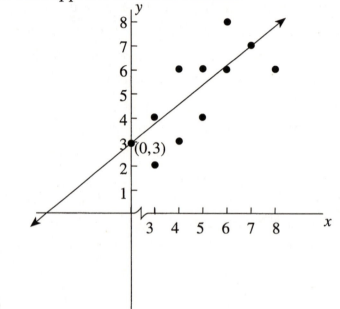

Figure 13–5:
The Modified
Form Of A
Scatter Diagram
With Incorrect
Y Intercept

The y intercept appears to be positive 3 instead of the correct value of 1. The error is due to the break in the horizontal axis that was made to provide a more pleasing graph.

Returning to Figure 13–4, we find the slope by selecting two points on the line. It helps if the two points are actual ordered pairs from our bivariate data set that are on the line. We then use the slope formula from algebra to find the value of b.

$$b = \frac{y_2 - y_1}{x_2 - x_1}$$

We will use the ordered pairs $(0,1)$ and $(7,7)$ to find the slope.

$$b = \frac{7-1}{7-0} = \frac{6}{7} \doteq 0.857$$

We now have the approximate values for the slope and y intercept. The equation of our estimated line of least squares is

$$y = 0.857x + 1$$

As a comparison, we can get the equation of the line using the principle of least squares. We can use the equations in the text for the slope and intercept of the regression line, or we can use the stat package in the HP–21S. Either way, we get the following equation:

$$\hat{y} = 0.795x + 1.145$$

Our estimate was really quite good. We do not always do this well, but we can at least get an idea of whether or not the values we calculate using the equations for the slope and intercept are reasonable. Even if we use a calculator to get the equation of the regression line, it is important to have a visual sense of what the line should look like before we enter the data into the calculator.

Hot Spot #2 Sample Problem and Solution: Given the following graph of a line, find the slope and y intercept. Then write the equation of the line in the slope-intercept form $y = a + bx$.

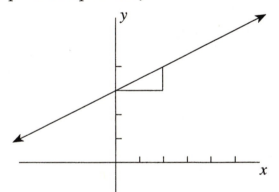

Figure 13–6:
Finding Slope
And Y Intercept
From A Function
Graph

Solution: Step 1.   The line crosses the y axis at $(0,3)$.

Step 2.   The slope of the line is $\frac{1}{2}$.

Step 3.   The equation in slope-intercept form is $y = \frac{1}{2}x + 3$.

Sample Problem and Answer: Given the following graph of a line, find the slope and y intercept. Then write the equation of the line in the slope-intercept form $y = a + bx$.

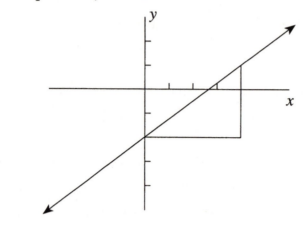

Figure 13–7:
More Finding Of
Slope And
Y Intercept From
A Function
Graph

Answer: $b = \frac{3}{4}$, $a = -2$, $y = \frac{3}{4}x - 2$

Sample Problem and Answer: Given the following graph of a line, find the slope and y intercept. Then write the equation of the line in the slope-intercept form $y = a + bx$.

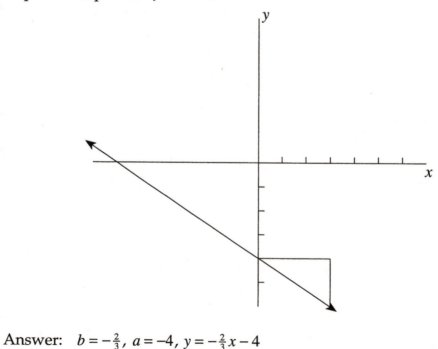

Figure 13–8: Yet
More Finding Of
Slope And
Y Intercept From
A Function
Graph

Answer: $b = -\frac{2}{3}$, $a = -4$, $y = -\frac{2}{3}x - 4$

Hot Spot #3 – The Effect Of Errors On The Regression Line And The Correlation Coefficient

✳
Mann Section 13.2.2

The formulas we use for finding the slope, b, and y intercept, a, of the line of least squares and the correlation coefficient, r, are very sensitive to errors in the summations we use. This is particularly true if the data values are small. The following set of bivariate data will be used to show what happens when errors are made:

x	y	x^2	y^2	xy
1	1.2	1	1.44	1.2
1.5	0.8	2.25	0.64	1.2
2	0.6	4	0.36	1.2
3	0.4	9	0.16	1.2
4	0.3	16	0.09	1.2

Table 13–1: Table of Sample Bi-Variate Data

The totals for each column are found as follows:
$$\sum x = 11.5 \quad \sum y = 3.3 \quad \sum x^2 = 32.25 \quad \sum y^2 = 2.69 \quad \sum xy = 6$$

When the above sums are substituted into the formulas for b, a, and r, we get the following values:
$$b = -0.27413 \qquad a = 1.2905 \qquad r = -0.9227$$

We are now ready to see what happens if an error is made in one of the entries in the table. One of the most common mistakes is making an error in squaring one of the values in the table. The values of x and y in the above table are easy to square, but let's see what happens when we make an error in squaring $y = 0.3$. If $(0.3)^2$ were written in error as 0.9, the only sum that would change is Σy^2, which would increase to 3.5. In this case the values of b and a do not change because the formulas do not use Σy^2. But there is a drastic change in the r value. The new value of r is -0.574205.

If we look at the value of r^2 for the two different values of r we obtained, we can really see the effect of the error. The r^2 of 0.8514 for the correct value of r tells us that approximately 85% of the variation in the y values is due to the regression line. When the error is made, the new r^2 is approximately 33%, which is a reduction of more than 50% in explained variation.

Now let's look at errors that also affect values of b and a. If the error were made in a value of x^2, the error changes the value of the denominator in the formula for both b and r and thus changes the value of all three computations. We can illustrate this last point by looking at what happens if the square of 1.5 is written incorrectly as 22.5 rather than 2.25. The Σx^2 is increased by 20.25 and the effect on b, a, and r is again drastic. The new values based on the error are as follows:

$$b = -0.0610 \qquad a = 0.8004 \qquad r = -0.4354$$

Of the two errors we examined this last one is the more serious because all three values are affected to such a degree that we might as well throw out the entire analysis. Again the point is that care must be taken in the calculations used in the formulas for b, a, and r.

The obvious solution to this problem of errors is to use a calculator or computer to obtain the values for the regression line and the correlation coefficient. If we must do the calculations using the formulas, it is imperative that we avoid errors in the table of values we create to find b, a, and r.

Hot Spot #4 – Using A t Test Statistic For A Hypothesis Test Of $H_o: B = 0$ Versus $H_a: B \neq 0$

Mann
Section 13.5.3

The hypothesis test for the null hypothesis $H_o: B = 0$ fits the template model introduced earlier in chapter 9 (page 9–5) where

$$z = \frac{[] - \mu_{[]}}{\sigma_{[]}}, \text{ where the [] can be replaced by}$$

1. x
2. $\bar{x}$
3. $\hat{p}$
4. $\bar{x}_2 - \bar{x}_1$
5. $\bar{d}$

Template For z
Test Statistic

We can now add the sample statistic b to our list:

6. b

In the case of b, when $B = 0$, the sampling distribution for small samples with $n < 30$ involves a t test with $n - 2$ degrees of freedom. We lose the 2 degrees of freedom because we need to calculate both $\bar{x}$ and $\bar{y}$.

Returning to our template, we need to replace z with t. When we place b in the box, the t test looks like the following:

$$t = \frac{b - \mu_b}{\sigma_b} \text{ where } \mu_b = B \text{ and } \sigma_b = \frac{\sigma_e}{\sqrt{SS_{xx}}} \text{ and } s_b = \frac{s_e}{\sqrt{SS_{xx}}}$$

The final form of our test statistic is:

$$t = \frac{b - B}{s_b} \text{ with } df = n - 2$$

Returning to the hypothesis test of $B = 0$, once we have a value of b and s_b, the t test is easy to do. We follow the same five steps we have been using for hypothesis testing in the last three chapters.

Step 1. State the hypotheses.
Step 2. Select the distribution to use.
Step 3. Determine the rejection region.
Step 4. Find the value of the test statistic.
Step 5. Either reject or fail to reject the null hypothesis.

Hot Spot #4 Sample Problem and Solution: Given $b = 0.3$, $s_b = 0.225$, and $n = 20$, test $H_o: B = 0$ and $H_a: B \neq 0$ at the 5% level of significance.

Answer: Step 1. $H_o: B = 0$ and $H_a: B \neq 0$.

Step 2. $n < 30$ and s_b is used for σ_b, so we use t.

Step 3. The critical value of t for a two-tailed test at $df = 18$ is $t = \pm 2.101$.

Step 4. $t = \dfrac{b - B}{s_b} = \dfrac{0.3 - 0}{0.225} = 1.33$

Step 5. Fail to reject.

Sample Problem and Answer: Given $b = 0.4$, $s_b = 0.173$, and $n = 29$ test $H_o: B = 0$ and $H_a: B > 0$ for significance at the 5% level.

Answer: Step 1. $H_o: B = 0$ and $H_a: B > 0$.

Step 2. $n < 30$ and s_b is used for σ_b, so we use t.

Step 3. The critical value of t for a one-tailed test at $df = 27$ is $t = 1.703$.

Step 4. $t = \dfrac{b-B}{s_b} = \dfrac{0.4-0}{0.173} = 2.31$

Step 5. Reject H_0 in favor of the alternative $H_a : B > 0$.

Sample Problem and Answer: Given $b = 0.2$, $s_b = 0.099$, and $n = 100$ test $H_o : B = 0$ and $H_a : B \neq 0$ for significance at the 5% level.

Answer: Step 1. $H_o : B = 0$ and $H_a : B \neq 0$.

Step 2. $n > 30$, so we use the normal distribution for the critical value of t.

Step 3. The critical value of t for a two-tailed test at $df = 98$ is $t = \pm 1.96$.

Step 4. $t = \dfrac{b-B}{s_b} = \dfrac{0.2-0}{0.099} = 2.02$

Step 5. Reject H_0 in favor of the alternative $H_a : B \neq 0$.

Hot Spot #5 – The Danger Of Using The Regression Line To Predict Values Of y For x Outside The Interval Used To Obtain The Line

✳
*Mann
Sections 13.3,
13.2.5, And 13.9*

Many times the scatter diagrams appear to be clustered around a line when in fact the relationship between the variables is curvilinear. The apparent linear relationship may be because the values of x are restricted to a particular interval. The equation $xy = 12$, for x and y both positive, gives a good illustration of the problems that can exist. The graph of the equation is a hyperbola in the first quadrant as shown below, in Figure 13–9:

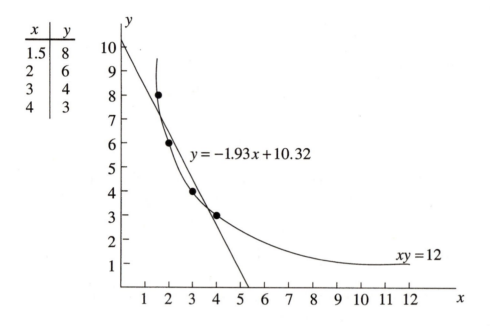

Figure 12–9:
A Scatter
Diagram Of
Values On The
Curvi-Linear
Hyperbola
xy=12

If the bivariate set of data is limited to values of x between 1 and 4 as illustrated in the above data set, the graph appears to cluster around the regression line $y = -1.932x + 10.322$ that is found using the 4 ordered pairs in the table. If we use the regression line to find the value of y when x is 10, we get

$$y = -1.932(10) + 10.322 = -9$$

It is clear from the graph of $xy = 12$ that the y value must always be positive. The correct value for y is found by substituting 10 into the equation $xy = 12$.

$$y = \tfrac{12}{10} = 1.2$$

Yet when we use the regression line we get a negative value of y when $x = 10$. The residual e is

$$e = 1.2 - (-9) = 10.2$$

The error is 850% ($\tfrac{10.2}{1.2}$). The problem is that the linear relationship we created with the regression line was only valid for the interval from 1 through 4. When we use the regression line to find y for $x = 10$ we are assuming that the linear relationship is still valid. In this example we made a large error in saying $y = -9$ when $x = 10$. As a rule of thumb, we should stay within 20% of the width of our original interval. In the above example the width of the original interval is 3 and 20% of 3 is 0.6.

We can now see how close we are to the correct value of y when we use the largest value of x allowed, that is, $4 + (0.02)(3) = 4.6$:

$$y = -1.932(4.6) + 10.322 = 1.435$$

The correct value of y is found by substituting 4.6 into the equation $xy = 12$.

$$y = \tfrac{12}{4.6} = 2.609$$

We still have an error, in that the residual value e is

$$e = 1.435 - 2.609 = -1.174$$

The error of 45% $\left(\tfrac{1.174}{2.609}\right)$ is large; but when compared to the error of 850% that we had using $x = 10$, we have a considerable improvement. The point is that we will have an error in either case. But if we limit ourselves to 20% of the width of the original interval for x, the error will at least be reasonable.

Of course, in practice we never know if the bivariate data set we use for linear regression actually belongs to a curvilinear relation. Relative to the interval $(1, 4)$ we used to get the regression line, the correlation coefficient $r = 0.966$. When we square r, we get the coefficient of determination r^2. This value is 0.933 and it tells us that 93% of the variation in y is explained by the regression. The point is that relative to our interval of $(1, 4)$ the regression line is a good model that explains the clustering we have about the line. It is only when we go outside of the interval that we have the potential for making large errors in estimating y. Of course, if our correlation coefficient is very small we may still make large errors in estimating y independent of whether or not there is a curvilinear relationship. Indeed it is possible to have bivariate data sets that exhibit no relationship whatsoever.

CHAPTER 13 DISCUSSION QUESTIONS

These questions may be used in your study group or simply as topics for individual reflection. Whichever you do, take time to explain verbally each topic to insure your own understanding. Since these questions are intended as topics for discussion, answers to these questions are not provided. If you find that you are not comfortable with either your answers or that your group has difficulty with the topic, take time to meet with your professor to get help.

1. How can you use the graph of the bivariate data to determine if the correlation coefficient is positive or negative?

2. How can you use the scatter diagram to get approximate values of the slope and y intercept for the regression equation?

3. How is the coefficient of determination used to evaluate the degree of linear relationship between two variables?

4. What is the t test statistic that is used with hypothesis testing of $H_o:B = 0$ versus $H_a:B \neq 0$?

5. Why does the interval estimate of y change as the values of x are further away from the mean of x?

6. Why is a linear relationship between two variables not necessarily a causal relationship?

7. How is it possible that a weak linear relationship between two variables can result in a significant value of b in a hypothesis test of $H_o:B = 0$?

8. How do errors in the table used for calculations affect both regression and correlation analysis?

9. What are the assumptions of the linear regression model?

10. How might it be possible that two variables are related even though the linear coefficient is nearly zero?

CHAPTER 13 TEST

Use the following information for problems 1, 2 and 3:

$$\sum x = 10 \qquad \sum y = 14 \qquad \sum xy = 58 \qquad \sum x^2 = 44 \qquad \sum y^2 = 80 \qquad n = 4$$

1. Find b and a.

2. Find r.

3. Find r^2.

4. Given $b = 0.3$, $s_b = 0.138$, and $n = 50$, use a t test to do a hypothesis test on $H_o : B = 0$ versus $H_a : B \neq 0$ at $\alpha = 0.05$.

5. Given $\sum (y - \bar{y})^2 = 122.5$ and $\sum e^2 = 43.2$, find r^2.

6. Given the following graph of $y = a + bx$, estimate b and a:

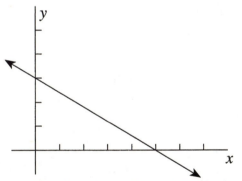

7. Given the regression line $\hat{y} = 1.25x + 4.5$, find a point estimate for y when $x = 1.5$.

8. Given that the ordered pair $(1.5, 6.1)$ belongs to the bivariate data set, find the error for the estimate in problem 7.

Use the following information for problems 9 and 10:

$$SST = 0.235 \qquad SSE = 0.025 \qquad SSR = 0.21 \qquad n = 10$$

9. Calculate s_e.

10. Calculate r^2.

CHAPTER 13 TEST Questions and Answers

Use the following information for problems 1, 2 and 3:

$$\sum x = 10 \qquad \sum y = 14 \qquad \sum xy = 58 \qquad \sum x^2 = 44 \qquad \sum y^2 = 80 \qquad n = 4$$

1. Find b and a.

Answer: $b = \dfrac{\left(\sum xy\right) - \dfrac{\left(\sum x\right)\left(\sum y\right)}{n}}{\left(\sum x^2\right) - \dfrac{\left(\sum x\right)^2}{n}}$

$$b = \dfrac{58 - \dfrac{(10)(14)}{4}}{44 - \dfrac{(10)^2}{4}} = \dfrac{92}{76} = 1.2105$$

$$\bar{x} = 2.5 \qquad \bar{y} = 3.5$$

$$a = \bar{y} - b\bar{x} = 3.5 - 3.026 = 0.4737$$

2. Find r.

Answer: $r = \dfrac{SS_{xy}}{\sqrt{SS_{xx}}\sqrt{SS_{yy}}} = \dfrac{\left(\sum xy\right) - \dfrac{\left(\sum x\right)\left(\sum y\right)}{n}}{\sqrt{\left(\sum x^2\right) - \dfrac{\left(\sum x\right)^2}{n}}\sqrt{\left(\sum y^2\right) - \dfrac{\left(\sum y\right)^2}{n}}}$

$$r = \dfrac{(58) - \dfrac{(10)(14)}{4}}{\sqrt{(44) - \dfrac{(10)^2}{4}}\sqrt{(80) - \dfrac{(14)^2}{4}}} = \dfrac{23}{\sqrt{19}\sqrt{31}} = 0.9477$$

A Study Guide For Statistics

3. Find r^2.

Answer: $r^2 = (0.9477)^2 = 0.8981$

4. Given $b = 0.3$, $s_b = 0.138$, and $n = 50$ use a t test to do a hypothesis test on $H_o: B = 0$ versus $H_a: B \neq 0$ at $\alpha = 0.05$.

Answer: $df = 48 \qquad t_{cv} = \pm 1.96 \qquad t = \dfrac{0.3}{0.138} = 2.18$

Conclusion: Reject H_0.

5. Given $\sum (y - \bar{y})^2 = 122.5$ and $\sum e^2 = 43.2$, find r^2.

Answer: $SSR = SST - SSE = 122.5 - 43.2 = 79.3$

$r^2 = \dfrac{SSR}{SST} = \dfrac{79.3}{122.5} = 0.65$

6. Given the following graph of $y = a + bx$, estimate b and a:

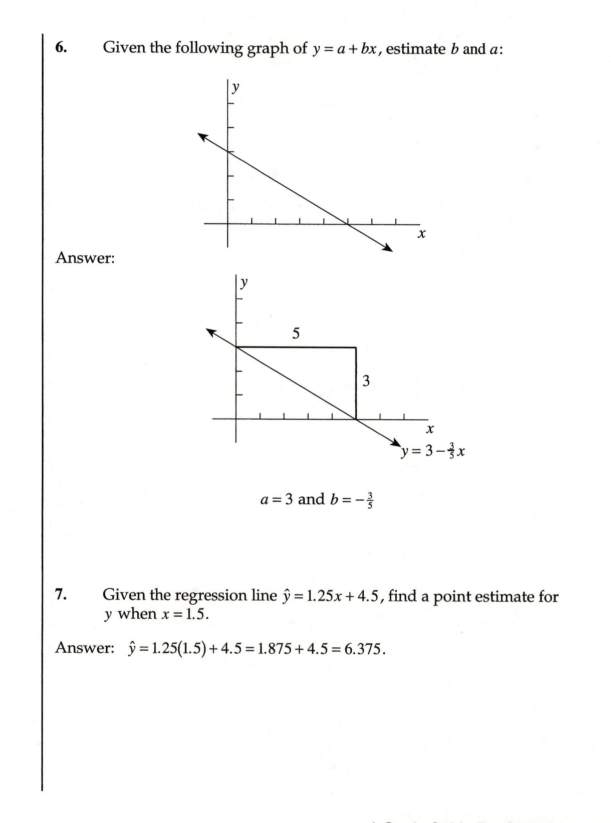

Answer:

$$a = 3 \text{ and } b = -\tfrac{3}{5}$$

7. Given the regression line $\hat{y} = 1.25x + 4.5$, find a point estimate for y when $x = 1.5$.

Answer: $\hat{y} = 1.25(1.5) + 4.5 = 1.875 + 4.5 = 6.375$.

8. Given that the ordered pair $(1.5, 6.1)$ belongs to the bivariate data set, find the error for the estimate in problem 7.

Answer: $e = \hat{y} - y = 6.375 - 6.1 = 0.275$.

Use the following information for problems 9 and 10:
$$SST = 0.235 \qquad SSE = 0.025 \qquad SSR = 0.21 \qquad n = 10$$

9. Calculate s_e.

Answer: $s_e = \sqrt{\dfrac{SSE}{n-2}} = \sqrt{\dfrac{0.025}{8}} = \sqrt{0.003125} = 0.055901699$.

10. Calculate r^2.

Answer: $r^2 = \dfrac{SSR}{SST} = \dfrac{0.21}{0.235} = 0.8936$.

$$\Sigma\circ$$